139.00

QP
376
.M37
2011

Miguel Marín-Padilla

The Human Brain

Prenatal Development and Structure

Author
Miguel Marín-Padilla
Timberlea Circle 2447
55125 Woodbury, MN
USA
marin4477@yahoo.com

ISBN 978-3-642-14723-4   e-ISBN 978-3-642-14724-1
DOI 10.1007/978-3-642-14724-1
Springer Heidelberg Dordrecht London New York

Library of Congress Control Number: 2010936628

© Springer-Verlag Berlin Heidelberg 2011
This work is subject to copyright. All rights are reserved, whether the whole or part of the material is concerned, specifically the rights of translation, reprinting, reuse of illustrations, recitation, broadcasting, reproduction on microfilm or in any other way, and storage in data banks. Duplication of this publication or parts thereof is permitted only under the provisions of the German Copyright Law of September 9, 1965, in its current version, and permission for use must always be obtained from Springer. Violations are liable to prosecution under the German Copyright Law.

The use of general descriptive names, registered names, trademarks, etc. in this publication does not imply, even in the absence of a specific statement, that such names are exempt from the relevant protective laws and regulations and therefore free for general use.

Product liability: The publishers cannot guarantee the accuracy of any information about dosage and application contained in this book. In every individual case the user must check such information by consulting the relevant literature.

Cover design: eStudioCalamar, Figueres/Berlin

Printed on acid-free paper

Springer is part of Springer Science+Business Media (www.springer.com)

I dedicated this monograph to the memory of my dear Teresa, spouse, mother of my children, friend and life-long companion, since student years. She understood my persona with compassion and let me to pursue, not always in focus and/or comprehensible, my efforts. Although, a mystery to me, I am deeply beholden to her. And also to the memory of Don Quijote, Cajal and Beethoven, inseparable companions on my itinerary and to young neuroscientists.

Woodbury, Minnesota 2010. *Miguel Marín-Padilla*

Preface

The present monograph explores the prenatal development of the human brain motor cortex from its emergence in the undifferentiated neuroepithelium, through the establishment of its primordial organization and the subsequent ascending and stratified cytoarchitectural and functional organizations. The prenatal development of the motor cortex basic neuronal, fibrillar, synaptic, microvascular, and neuroglial systems follows a concomitant, ascending and stratified progression from lower (older) to upper (younger) cortical strata. The monograph introduces new developmental data that explain, for the first time, why, how, and in what order the mammalian cerebral cortex becomes stratified (laminated) as well as the number of pyramidal cells strata that characterizes each mammalian species. It represents the first systematic exploration and description of the prenatal development maturations of the human motor cortex recorder. The rapid Golgi procedure has been used primarily in this developmental study with some additional staining procedures (Chapter 12). The monograph reassembles, organizes by age, discusses, and illustrates lifelong rapid Golgi studies of the developing brains of humans, cats, hamsters, and mice embryos and fetuses. New insights into the mammalian cerebral cortex developmental strategy and progressive ascending stratification (laminations) are discussed. A new developmental cytoarchitectonics theory and a new nomenclature, applicable to all mammalian species, are introduced. The new cytoarchitectonics theory proposes that the structural and functional maturations of the mammalian cerebral cortex are ascending and stratified processes that concomitantly involve its essential systems. The new theory challenges the current and universally held conception of descending (Layers I, II, III, IV, V, VI, VII) cortical laminations. To best describe some of the mammalian cerebral cortex developing fundamental features, new terms are introduced that should replace the less-specific ones currently used.

In separate chapters, the monograph describes and illustrates the following topics: Chapter 2, Mammalian Cerebral Cortex: Embryonic Development and Cytoarchitecture; Chapter 3, Human Motor Cortex: Development and Cytoarchitecture; Chapter 4, The Mammalian Pyramidal Neuron: Development, Structure and Function; Chapter 5, Human Motor Cortex First Lamina: Development and Cytoarchitecture; Chapter 6, Human Motor Cortex Excitatory–Inhibitory Neuronal Systems: Development and Cytoarchitecture; Chapter 7, Human Cerebral Cortex Intrinsic Microvascular System: Development and Cytoarchitecture; Chapter 8, Human Motor Cortex First Lamina and Gray Matter Special Astrocytes: Development and Cytoarchitecture; Chapter 9: New Developmental Cytoarchitectonics Theory and Nomenclature; Chapter 10, Epilogue; Charter 11, Cat Motor Cortex: Development and Cytoarchitecture (that corroborates many aspects of the developing human motor cortex).

The monograph Epilogue recapitulates the major developmental observations, explores their significance and implications, and introduces some personal insights into the possible anatomical substrate for those attributes that are uniquely human. The monograph emphasizes that in the course of mammalian ontogeny and phylogeny the number of cortical strata increases by adding new ones to those previously established. The number of cortical strata reflects the motor capabilities of each mammalian species. A separate appendix is dedicated to an analysis of the rapid Golgi procedure, from a personal perspective of many a year of using it (Chapter 12). This section aims at clarifying some of the misconceptions that surround this classic staining procedure and explains its capricious nature, extraordinary revealing capabilities, universal applicability, and authenticity. It emphasizes that to learn the rapid Golgi procedure and use it successfully requires perseverance and many years of dedication.

The monograph color illustrations celebrate the rapid Golgi procedure revealing potentials, beauty, clarity, and authenticity. Also its capability of staining neurons, fibers, synapses, blood capillaries (and growing capillaries), and glial elements, thus permitting to explore their variable morphologies and spatial (three-dimensional) interrelationships. The rapid Golgi color illustrations selected and reassembled in this monograph, reproduced herein for the first time, are unique and unavailable in either classic and/or current scientific literature. The monograph hopes to revive interest in this classic procedure and encourage young neuroscientists to learn and apply it in their studies.

Because of its enormous functional size, Golgi studies of the entire cerebral cortex are presently impossible. For such a study we will need: (a) a Brain Institute with dozen of committed investigators prepared to study the various regions of each brain received; (b) satellite Medical Centers to gather fresh (unspoiled) postmortem brains specimen; and (c) a system for bringing the material to the Brain Institute for processing and study, something similar to transporting fresh donor organs. Lacking these options, the best alternative – perhaps the only one – is to study a selected region of young brains, accumulate the data, make it available to guide additional studies of the same and/or other cortical areas. A single investigator, even if he/she devotes their entire life to it, can only study a small cortical area, preferably the same one and from different aged brains. Which I have done throughout my entire academic life and the observations made are reassembled and presented in this monograph.

The vertebrate central nervous system (CNS) might be conceived as a stratified series of nuclei composed of neurons that receive sensory information from the animal surroundings and transmit the information to motor neurons for controlling the animal motor activities in the searching for food, mate, and/or avoiding danger. In the course of vertebrate evolution, as new adaptations and novel motor activities emerge, new nuclei are established for their control. The new nuclei are added to previously established ones, in a caudal–cephalad progression along the animal neural axis (Llinás 2003). Each nucleus is also a stratified system composed of neurons that received sensory information from the animal surrounding, of neurons for controlling its motor activities, and of interneurons that interconnect sensory and motor neurons. Each nucleus develops its own intrinsic microvascular and neuroglial systems. During vertebrate evolution, the number of added nuclei increases progressively as well as their cytoarchitectural complexity. In the course of mammalian evolution, the cytoarchitectural complexity achieved by the added nuclei increases exponentially and, in general, remains poorly understood.

Preface

The last nucleus added to the vertebrate CNS is the mammalian cerebral cortex or neocortex. Throughout mammalian evolution, the neocortex remains a single, stratified and biologically open nucleus capable of further expansion by increasing the number of additional pyramidal cell strata. Understanding the neuronal, fibrillar, synaptic, microvascular, and neuroglial cytoarchitectural organization and composition of the entire cerebral cortex of any mammal is, at present, beyond our possibilities. Even understanding the organization of a single cortical region from an adult animal brain may be beyond our present resources. Consequently, the only option – as Cajal recommended – is to study the developing cerebral cortex with an appropriate neurohistological staining procedure capable of staining all of its elements. In the developing cerebral cortex, the basic elements are fewer, smaller, and easier to visualize, interpret, and understand. Even, a developmental study of a single cortical region will require a life-long dedication, accepting the impossibility of understanding its complete cytoarchitectural and functional organizations and much less that of proximal and distal cortical regions functionally interconnected with it. Despite these shortcoming, the basic neuronal, fibrillar, synaptic, microvascular, and glial systems of a single cortical area can be studied, though partially, with the rapid Golgi procedure, if sufficient amount of time and perseverance are dedicated to it, as the present monograph hopes to prove.

The information presented in the monograph pertains almost exclusively to the prenatal development of the human motor cortex. Despite 36 years of studying this relatively small area and over 4,000 preparations, what I have been able to observe and learn is merely a fraction of its overall cytoarchitectural organization. Despite its incompleteness, the developmental rapid Golgi observations, presented in this monograph, should provide a general view of the developing human motor cortex basic components and of their interrelationships. Comparative developmental Golgi studies of the motor cortex of cats, hamsters, and mice embryos and fetuses provide additional and corroborating information concerning the mammalian cerebral cortex developmental strategy. The monograph introduces three basic propositions: (a) The mammalian cerebral cortex prenatal development and functional maturation is an ascending, progressive and stratified process from lower (older) to upper (younger) strata that concomitantly involve all its essential components; (b) The number of pyramidal cells strata established in the cerebral cortex varies among different mammalian species; and (c) The number of pyramidal cell functional strata formed in the cerebral cortex reflects the animal motor capabilities and may distinguish the motor cortex of each mammalian species.

Since what the brain actually does is to move things, essentially muscles, the following working hypothesis was considered in selecting the study of the human brain primary motor cortex. Information from different cortical regions is channeled to the motor cortex for action or execution. The mammalian cerebral cortex essential functional outlet is through the motor cortex large pyramidal neurons. The assumption was that the motor capabilities of any mammalian species should be reflected on the cytoarchitectural organization and complexity of its primary motor cortex, such that the mouse motor cortex cytoarchitectural organization should be simpler (quantitatively and qualitatively) than that of humans and that the organization of the cat motor cortex should lie somewhere between them. In other words, a mouse, to accomplish all its motor needs, will need a cytoarchitectural organization in its motor cortex that should reflect its inherent and/or acquired motor capabilities. Similarly, the cytoarchitectural organization of a cat motor cortex should reflect its motor capabilities and should be more complex than that of a mouse and simpler than that of a primate motor cortex. Consequently, the cytoarchitectural organization of the human motor cortex should reflect his extraordinary – inherent and learned – motor capabilities (including motor creativity) and should be more complex than that of other primates. These differences might be expressed in the number of functional pyramidal cell strata established in the motor cortex of each mammalian species. The present monograph explores and reaffirms these assumptions. If this working hypothesis turns out to be correct, the universally held idea (since Darwin and classic neuroanatomical studies) that the only cytoarchitectural differences in the cerebral cortex among mammals is one of degree and not of kind should be reevaluated.

Woodbury, Minnesota, 2010   Miguel Marín-Padilla

Contents

1 Introduction 1

2 Mammalian Cerebral Cortex: Embryonic
  Development and Cytoarchitecture 5

3 Human Motor Cortex: Development and Cytoarchitecture 11
  3.1 Seven-Week-Old Stage 14
  3.2 Eleven-Week-Old Stage 16
  3.3 Fifteen-Week-Old Stage 18
  3.4 Twenty-Week-Old Stage 25
  3.5 Thirty-Week-Old Stage 26
  3.6 Thirty-Five-Week-Old Stage 27
  3.7 Forty-Week-Old (Newborn) Stage 29

4 The Mammalian Pyramidal Neuron: Development,
  Structure, and Function 35
  4.1 Developmental Features 35
  4.2 Morphological Features 38
  4.3 Functional Features 39
  4.4 Ascending Functional Maturation 43
  4.5 Descending Function 45

5 Human Motor Cortex First Lamina: Development
  and Cytoarchitecture 49
  5.1 First Lamina Principal Components 53
  5.2 Cajal–Retzius Cell Unique Morphology 55
  5.3 First Lamina Secondary Components 62

6 Human Motor Cortex Excitatory–Inhibitory Neuronal Systems:
  Development and Cytoarchitecture 65
  6.1 The Pyramidal-Basket System 66
  6.2 The Pyramidal-Martinotti System 73
  6.3 The Pyramidal-Double-Bouquet System 75
  6.4 The Pyramidal-Chandelier System 77

7 Human Cerebral Cortex Intrinsic Microvascular System: Development and Cytoarchitecture ... 85

8 Human Motor Cortex First Lamina and Gray Matter Special Astrocytes: Development and Cytoarchitecture ... 101
- 8.1 First Lamina Special Astrocytes ... 104
- 8.2 Grey Matter Protoplasmic Astrocytes ... 104

9 New Developmental Cytoarchitectonic Theory and Nomenclature ... 113

10 Epilogue ... 121

11 Cat Motor Cortex: Development and Cytoarchitecture ... 125
- 11.1 Cat 25-Day-Old Stage ... 125
- 11.2 Cat 30-Day-Old Stage ... 125
- 11.3 Cat 35-Day-Old Stage ... 127
- 11.4 Cat 40-Day-Old Stage ... 128
- 11.5 Cat 45-Day-Old Stage ... 129
- 11.6 Cat 50-Day-Old Stage ... 130
- 11.7 Cat 55-Day-Old Stage ... 131
- 11.8 Newborn Cat Motor Cortex ... 132

12 The Rapid Golgi Reaction: A Personal Quest ... 137

Index ... 143

About the Author ... 145

Introduction 1

Why is the mammalian cerebral cortex, including that of man, stratified (laminated) and how many cortical strata are established on the different species are fundamental questions not yet adequately addressed and/or answered. Others have expressed similar concerns: "We have no idea why the histology of the cerebral cortex is what it is" (Braitenberg 1978), "Laminations is one of the hallmarks of the cerebral cortex. Despite the early recognition of this characteristic feature of cortical histology, however, there is still little understanding of the fundamental organizing principle that governs it" (Jones, 1984), "It is obviously not possible in the present state of the science to define the determinants of cortical cytoarchitecture" (Jones 1990), "We have yet to achieve a universal theory of cortical organization" (Rakic and Singer 1988). This monograph introduces developmental data that explains, for the first time, why, how, and in what order the mammalian cerebral cortex becomes stratified and about the number of strata that characterizes each species. It also represents the first systematic study of the developing human cerebral cortex using primarily a single method, namely, the rapid Golgi procedure.

While we understand, in great detail, the structural organization of any organ and/or tissue of the human body, that of our brain has remained surprisingly poorly understood. Moreover, the idea that the only cytoarchitectural difference in the cerebral cortex among mammals is "one of degree and not of kind" has prevailed and is universally accepted (Darwin 1936). This old Darwinian idea, which remains unsupported by developmental data, need to be reevaluated. Although this monograph addresses some of these questions, it does not pretend to be a review of all the information available concerning the mammalian cerebral cortex overall structural and functional organizations, which already have been successfully reassembled and described (Creutzfeldt 1995). The monograph essentially describes the prenatal developmental histories of the human motor cortex gray matter basic neuronal, synaptic, excitatory-inhibitory, intrinsic microvascular, and neuroglial systems.

Only a developmental study of the mammalian cerebral cortex will provide the necessary information for clarifying why, how, and in what order is its cytoarchitecture stratified and/or laminated. Not until we understand how the mammalian cerebral cortex becomes progressively stratified we shall remain ignorant about its basic structural organization and unable of adequately explaining its function. And also unable of answering what types of alterations affect the cerebral cortex multiple systems on either genetic and/or acquired neurological and psychological disorders. The monograph also introduces rapid Golgi observations on the prenatal development of the cat motor cortex, which corroborate the human data.

Recent technological advances in the fields of brain-scanning, neurogenetics, neurobiology, behavioral neurosciences, and molecular biology, to mention a few, have displaced, to a distant second place, human brain cytoarchitectural studies. Perhaps, these neurotechnological advances (often spectacular) are conveying the general impression that they are answering and/or resolving much more than they really do and/or resolve. Without minimizing their merits and significant accomplishments, they have not yet provided us with a better understanding of the human brain basic cytoarchitectural organization. While an increase in blood flow throughout a brain region indicates that its neurons are actively functioning, it offers no information concerning the type of neurons involved and/or about their architectural and functional organizations. Moreover, a better understanding of the human brain structural organization should improve our understanding of brain-scanning data and, specially, its interpretation.

Despite these technological advances, there are important questions about the human brain that remain unanswered and/or unexplained. We continue to ignore its evolving cytoarchitectural organization through prenatal development, infancy, childhood and adulthood; we ignore what happen to the language center neurons when a child begins to speak or to those of the motor cortex when he or she learns new motor activities (writings, painting, sculpturing, playing piano, violin, or any sport); we ignore how the frontal lobe neurons, where information is reassembled, intuited, and sent for action, evolve and are organized; we ignore how the neurons of a damaged brain recover, reorganize, and reestablish new (possibly altered) proximal and distal functional interconnections; we ignored how an epileptic focus neurons evolve, reorganize, reestablish – proximal and distal – interconnections and if they are capable of modifying the post-injury reorganization of distant neuronal systems functionally interconnected with the original focus (perhaps explaining the spreading of seizures from one hemisphere to the other); we ignored what happen to the neurons in an autistic brain; and, despite all that has been said, written, and published we continue to ignore the basic neuronal, fibrillar, synaptic, interneuronal, microvascular, and neuroglial organization of most – genetic and/or acquired – neurological and psychological disorders. Perhaps, some of these incognitos will be better understood if the human brain basic cytoarchitectural organization will be better comprehended. This monograph offers new insights into the human brain essential structural organization.

Every establishment dedicated to the caring of the unborn, the prematurely born, and of infants recovering from perinatal brain damage should have charts and tables depicting the infant brain progressive development. The Neonatal Intensive Care Unit's (NICU) nurses, neonatologists, and pediatrician should be aware of the developmental maturation of the brain of infants under their care. They must understand which neurons are already developed and functioning and which ones are not and, for that reason, more vulnerable. All NICU and Pediatric Departments should be provided with developmental charts depicting the human brain development. This monograph provides the necessary data for the making of those charts and maps reflecting the human brain progressive cytoarchitectural development through different gestational ages. To improve the clinical management of perinatal brain damage and perhaps to better control its, often devastating, clinical outcome is also crucial to correlate the infant's age, at the time of injury, with the degree of his/her brain cytoarchitectural and functional maturations. The main emphasis on the clinical care of infants born prematurely has been devoted (with great success) to understanding the developing human respiratory system and to control its difficulties. It is time to consider the developmental maturation of the infant brains and start investigating ways of protecting it from perinatal hypoxia. Unfortunately, the incidence of neurological and psychological problems as well as poor school performance of children born prematurely (with supposedly minimal brain damage) continued to be (through the years) statistically significant (Marín-Padilla 1996, 1997, 1999, 2000; Marín-Padilla et al. 2002; Robertson et al. 1989). This book provides developmental brain data that could hopefully improve the clinical care of infant with perinatal brain damage and perhaps in preventing and/or ameliorating its devastating effects.

There are also young neuroscientists, the word-over, which an inadequate understanding on the mammalian cerebral cortex cytoarchitectural organization, this monograph should be of help to them. Nowadays, few developmental structural studies are being carried out (and/or supported) and few investigators are dedicated to its teaching. The monograph could also be a valuable tool for instructing medical students and other medical professionals. It also provides valuable information to neurologists, psychologists, psychiatrics, and pediatric neuropathologists and, in general, to basic, clinical, and/or experimental neuroscientists.

Concerning the monograph objectives, the following comments seem pertinent. An adult human brain functional size is unimaginable with trillions of functional interconnections, continuously evolving, adapting, and changing from billions of neurons. At the present time, its cytoarchitectural complexity is beyond comprehension and impossible to be adequately studied. Cajal suggested instead the study of the developing (the young forest) brain (Cajal 1923). Its structure is simpler, has fewer elements, is easier to understand, and may provide a clearer assessment of its various elements as well as their evolving structuro-functional interrelationships. The brain complex cytoarchitectural organization will be easier to comprehend by starting from its earlier and simpler organization and advanced sequentially to older and more complex ones. This monograph follows this approach and provides an assessment of the prenatal cytoarchitectural ascending and stratified organization of the human brain basic components.

To adequately study the developing human brain (or that of any other mammal), two indispensable prerequisites are required, namely, a direct access to unspoiled postmortem brain tissue and the use of a technique capable of staining its neuronal, fibrillar, synaptic, microvascular, and glial components. In addition, the human brain postmortem deterioration is a fast process, which in a few (2 to 3) hours most of its delicate neuronal, axonal, synaptic, microvascular, and glial features have deteriorated beyond an accurate recognition. This single drawback may be the central reason for explaining our persistent ignorance about its fine cytoarchitectural organization. Formalin-fixed postmortem brain preparations, which are universally used, are useless in demonstrating the neurons three-dimensional organization and interrelationships with other cortical components. These preparations, which only stain neuronal bodies, fail to demonstrate the extent and distribution of neuron's dendrites and axonic profiles, as well as the extent of their functional territories (compare a and b of Fig. 3.14 of Chap. 3). Therefore, the need to study fresh unspoiled postmortem brain tissue becomes an imperative, something that is seldom accomplished (Chapter 12).

To have access to unspoiled postmortem brain tissue requires a unique set of unlikely circumstances. A classic example is Cajal's original Golgi studies of the visual, motor, sensory, acoustic regions of newborn infant brains, published between 1899 and 1900. Cajal's original Golgi studies, already more than 110 years old, continue to be a unique source of information (Cajal 1899a, b, 1900a, b). Despite their age, the original Golgi preparations have retained their original transparency, clarity, beauty, and authenticity (see Fig. 3.12b of Chap. 3). These Golgi studies were carried out during his tenure as Medical Director of a Madrid Orphanage, position that, unfortunately, lasted only a year (see foot note [1]). Since Cajal's early studies, there have been very few additional systematic Golgi studies of the developing human brain.

After a fellowship (1966–1967) at the Cajal Institute (Madrid, Spain) to study Cajal's original Golgi preparations and familiarize myself with the procedure's capricious nature and capabilities, I returned to Dartmouth. There, I began using the rapid Golgi procedure to study the developing brain of humans, cats, hamsters, and mice embryos and fetuses (Chapter 12). A set of circumstances made this study possible. As Professor of Pathology and of Pediatrics and Director of the Pediatrics Autopsy Service at Dartmouth-Hitchcock Medical Center (1962–1999), I was familiar with most pediatric cases and had direct access to all cases (embryos, fetuses, premature infants, newborns, and young infants), which for a variety of medical problems, came for postmortem evaluation. With the understanding, approval, and collaboration of colleagues, nurses and, particularly, of parents, I was able to gather fresh unspoiled postmortem brain tissue as soon as it was possible and made thousands of rapid Golgi preparations (Chapter 12). All the Institution ethical and legal requirements were fulfilled.

The rapid Golgi procedure, (Golgi 1873) despite its antiquity, remains the best available technique for studying the developing nervous tissue (Cajal 1904, 1923; Marín-Padilla, 1971, 1972, 1978). This classic procedure, seldom used due to misconceptions and misunderstanding, continues to be the best one for staining the neuron's dendritic and axonic profiles, as well as their associated fibrillar, synaptic, intrinsic microvascular and neuroglial systems (Chaps. 3–8). In addition, the thickness of a rapid Golgi preparation (between 150 and 250 μm) permits unique three-dimensional views of the brain various components spatial interrelationships, and multiplying the data each section provides. During 36 consecutive years, I studied these rapid Golgi preparations and the

---

[1] I spent a year (1966–1967) at the Cajal Institute in Madrid, under a US National Institute of Health, Neurohistology Fellowship, to study Cajal's original Golgi preparations and learn the method. During that year, I met Professor Fernando de Castro (Cajal's last disciple), already quite ill, and we become friends. He used to come to his office, some evenings, and seating on his armchair would talk to us (a couple of interested fellows) about Cajal's work and life. Among many anecdotes, he told me that the pressing need to study the human brain moved Cajal to accept the position of Medical Director of a Madrid Orphanage, assuming the medical care of its inmates. The satisfied Sisters would call on him when medical needs arose, for the care of infants, for the signing death certificates and, occasionally, to perform postmortem evaluations. From these autopsies he obtained the fresh human brain tissue for his classic Golgi studies of the sensory, motor, visual, acoustic, and olfactory cortex of newborn infants. Unfortunately, after one year, he lost this position because of a religious disagreement with the Sisters, who failed to understand the significance of his work. His original Golgi papers of newborn infants cerebral cortex were published during that year (1899–1900). Professor De Castro encouraged me to follow Cajal footsteps and to study the developing human brain with the rapid Golgi procedure. I still cherish the gift of his personal porcelain plaque (# 79), representing Cajal's visage in a profile, sculptured by Victorio Macho, commemorating Cajal's first centenary. Back in Dartmouth Medical School, following his encouraging words and recommendations, I learn to manipulate the rapid variant of the Golgi procedure and persistently applying it to the study of the developing cerebral cortex of humans, cats, hamsters, and mice embryos and fetuses (Chapter 12). The results of these life-long studies are reassembled and amply illustrated in this monograph.

observations made, are scattered through 28 abstracts, 44 peer review papers, and 12 book chapters. These Golgi observations are reassembled and amply illustrated in this monograph.

In view of the magnitude of the project, this monograph only describes the prenatal development of a small region of the human brain, its motor cortex, and not even entirely but only a portion of it. The prenatal development of other regions of the human brain remains essentially unexplored. To gather the necessary material and to study any other cortical regions, with the rapid Golgi procedure, will require a similar amount of time and commitment. Consequently, the rapid Golgi observations presented in the monograph are still incomplete and should be considered as stepping-stones for future and must-need additional Golgi studies of the developing human brain. Throughout the monograph illustrations, the beauty, clarity, and authenticity of rapid Golgi preparations are reaffirmed and their extraordinary revealing capabilities homage. The monograph hopes to encourage young neuroscientists to learn the rapid Golgi procedure, select a region of the human brain (or from any other mammal), and dedicate part of their lives to its study.

These Golgi studies were supported (1966–2003) by grants from the United States National Institutes of Health (Child Health and Human Development and Neurological Disorders and Stroke). I was also honored (1989) with a Jacob Javits Neuroscientist Investigator Award, in recognition of my Golgi studies, which extended my research efforts for several years.

I extend my gratitude to Dartmouth Medical School, to my colleagues, and to all parents and pediatric nurses for allowing me to carry out Golgi studies on infant postmortem brains and to the United States National Institute of Health for supporting them.

References

Braitenberg V (1978) Cortical architectonics general and aereal. In: Brazier MAB, Petsche H (eds) Architectonics of the cerebral cortex. Raven, New York, pp 443–465

Cajal SR (1899a) Estudios sobre la corteza cerebral humana. Estructura de la corteza motríz del hombre y mamíferos. Revista Trimestrar Micrográfica 4:117–200

Cajal SR (1899b) Estudios sobre la corteza cerebral humana. Corteza visual. Revista Trimestrar Micrográfica 4:1–63

Cajal SR (1900a) Estudios sobre la corteza cerebral humana. Estructura de la corteza acústica. Revista Trimestrar Micrográfica 5:129–183

Cajal SR (1900b) Estudios sobre la corteza cerebral humana. Estructura de la corteza cerebral olfativa del hombre y mamíferos. Revista Trimestrar Micrográfica 6:1–150

Cajal SR (1904) Textura del systema nervioso del hombre y de los vertebrados. Librería Nicolas Moya, Madrid

Cajal SR (1923) Recuerdos de mi Vida. Librería Juan Pueyo, Madrid

Creutzfeldt OD (1995) Cortex Cerebri. Performance, structure and functional organization of the cortex. Oxford University Press, Oxford

Darwin C (1936) The descent of man. The Modern Library, New York

Golgi C (1873) Sulla sostanza grigia del cervello. Gazetta Medica Italiana Lombardica 6:244–246

Jones E. (1984) Laminar distribution of cortical efferents cells. In: Peter A, Jones E (eds) Cerebral Cortex, Vol 1. Plenum Press, New York, pp 521–553

Jones EG (1990) Determinants of the cytoarchitecture of the cerebral cortex. In: Edelman GM, Gall WE, Cowan WM (eds) Signal and sense: local and global order in perceptual maps. Wiley, New York, pp 3–49

Marín-Padilla M (1971) Early prenatal ontogenesis of the cerebral cortex (neocortex) of the cat (Felis domestica). A Golgi study. Part I. The primordial neocortical organization. Z Anat Entwickl-Gesch 134:117–145

Marín-Padilla M (1972) Prenatal ontogenetic history of the principal neurons of the neocortex of the cat (Felis domestica). A Golgi study. Part II. Developmental differences and their significances. Z Anat Entwickl-Gesch 136:125–142

Marín-Padilla M (1978) Dual origin of the mammalian neocortex and evolution of the cortical plate. Anat Embryol 152:109–126

Marín-Padilla M (1996) Developmental neuropathology and impact of perinatal brain damage. I. Hemorrhagic lesion of neocortex. J Neuropathol Exp Neurol 55:746–762

Marín-Padilla M (1997) Developmental neuropathology and impact of perinatal brain damage. II. White matter lesions of the neocortex. J Neuropathol Exp Neurol 56:219–235

Marín-Padilla M (1999) Developmental neuropathology and impact of perinatal brain damage. III. Gray matter lesions of the neocortex. J Neuropathol Exp Neurol 58:407–429

Marín-Padilla M (2000) Perinatal brain damage, cortical reorganization (acquired cortical dysplasia), and epilepsy. In: Williamson P, Siegel AM, Roberts VM, Vijav VM, Gazzaniga M (eds) Neocortical epilepsies. Lippincott, Philadelphia, pp 156–172

Marín-Padilla M, Parisi JE, Armstrong DL, Sargent SK, Kaplan JA (2002) Shaken infant syndrome: developmental neuropathology, progressive cortical dysplasia, and epilepsy. Acta Neuropathol 103:321–332

Rakic P, Singer W (1988) Introduction. In: Rakic P, Singer W (eds) Neurobiology of the neocortex, Report of the Dahlem workshop. Wiley, Berlin, pp 1–4

Robertson CMT, Finer NN, Grace MGA (1989) School performance of survival of neonatal encephalopathy associated with asphyxia at term. J Pediatr 114:753–760

Mammalian Cerebral Cortex: Embryonic Development and Cytoarchitecture

The prenatal developmental of the mammalian cerebral cortex, including that of humans, is characterized by two sequential and interrelated periods: an early embryonic and a late fetal one. The embryonic period is characterized by the establishment of a primordial cortical organization, which is common to all mammals and shares features with the primitive (primordial) cortex of amphibians and reptiles (Marín-Padilla 1971, 1978, 1983, 1992, 1998). Its establishment represents a prerequisite for the subsequent formation, development, and organization of the pyramidal cell plate (PCP), (cortical plate (CP) in current nomenclature), which represents a mammalian innovation. The original description of the dual origin and composition of the mammalian neocortex has been corroborated and, today, is universally accepted. The mammalian fetal period is characterized by the sequential, ascending, and stratified organization of the PCP, which represents the most distinguishing feature of the mammalian cerebral cortex. The term pyramidal cell plate (PCP), used in the present monograph, replaces the current unspecific term of cortical plate (CP) for various reasons. The early PCP is composed solely of pyramidal neurons and represents the mammalian neocortex's most distinguishing feature (Chapters 3 and 4). Other neuronal elements are incorporated into the PCP later in prenatal development (Chapter 6). The progressive ascending and stratified anatomical and functional maturations of the PCP from lower (older) to upper (younger) pyramidal neurons are concomitant with the ascending and also stratified incorporation and maturation of its associated fibrillar, synaptic, excitatory-inhibitory, microvascular, and neuroglial systems (Chapters 4, 6–8).

The mammalian neocortex embryonic period starts in the very young embryo, evolves rapidly and represents a transient cortical organization. The establishment, throughout the embryonic cerebrum, of the primordial cortical organization follows a ventral (proximal) to dorsal (distal) gradient that coincides with the advancing penetration of primordial corticipetal fibers and of migrating primordial neurons throughout the cortex subpial region. This period has been studied in hamsters, mice, and cats and human embryos using both the rapid Golgi and routine staining procedures. The ventral (proximal) to dorsal (distal) progression and its basic neuronal and fibrillar composition is similar in all embryos studied and is considered to be essentially analogous in all mammalian species (see Fig. 11.9).

The mammalian cerebral cortex primordial cytoarchitectural organization goes through four sequential stages, which, in order of appearance, are: the undifferentiated neuroepithelium (NE), the marginal zone (MZ), the primordial plexiform (PP), and the PCP appearance. These stages evolve rapidly and sequentially throughout the cerebrum subpial zone in a ventral (proximal) to dorsal (distal) gradient. In whole brain preparations, it is possible to observe the transformations of one stage into the next (Figs. 11.3a and 11.9). This developmental gradient, observable in the four stages, has been crucial for understanding the appearance, chronology, and interrelationships of the cortex early neuronal and fibrillar components. Although, the following descriptions emphasize each embryonic stage essential features, they should be understood as a continuum.

Neuroepithelium (NE) Stage. The cerebral cortex original neuroepithelium is composed of closely packed ependymal (ventricular) cells attached to both the ependymal and the pial surfaces. The neuroepithelial cells are closely packed and their bodies are found throughout all levels excepting at the zone immediately bellow the pial surface occupied by their terminal filaments (Fig. 2.1a, b). The neuroepithelial cells are attached to the pial surface by filaments with several terminal endfeet, which united by tight junctions, build the pia external glia limiting membrane (EGLM) and manufacture its basal lamina (Fig. 2.1a, b). The pial surface represents the inner (basal) surface of the original neuroectoderm and its basal lamina before its closure and subsequent transformation into a neural tube. These cells are also attached to the ependymal surface. Cells at various stages of mitoses are recognized, in a ventral to dorsal gradient, through the ependymal surface (Fig. 2.1b, arrow heads). These mitotic cells add new components to the cortex expanding neuroepithelium and, eventually, some of them will participate in the generations of both neuronal and glial cells precursors destined to the cortex developing white and gray matters.

The mammalian neocortex NE stage is recognized in 9-day-old hamster and mouse, 19-day-old cat and 40-day-old human embryos (Fig 3.1a).

Marginal Zone (MZ) Stage. The continuing growth and surface expansion of the cortex requires of a sustained incorporation of new radial glial endfeet and additional basal lamina material. The increasing number of radial glial terminal filaments conveys to the subpial zone a light fibrillar appearance known as the MZ (Fig. 2.1a, 20 days). The presence of arriving primordial corticipetal fibers also contributes to its fibrillar appearance. A few scattered neurons, lacking distinctive morphological features, start to be recognized through the developing cortex proximal region (Fig. 2.1a, 20 days). The early primordial neurons and corticipetal fibers arrive at the MZ from extracortical sources, via the CNS subpial region (Zecevic et al. 1999). Some of these early neurons seem to originate in the medial ganglionic eminence. The origin of the early arriving primordial corticipetal fibers remains unknown. It has been suggested that some of them could represent monoaminergic fibers from mesencephalic and/or early thalamic centers; their origin needs to be further investigated (Marín-Padilla and T. Marín-Padilla 1982; Zecevic et al. 1999).

The neocortex MZ stage is recognized in 10-day-old hamster and mouse, 20-day-old cat, and 43-day-old human embryos (Fig. 2.1a, 20 days and see Chapter 3).

Primordial Plexiform (PP) Stage. As the number of primordial corticipetal fibers and of neurons increases throughout the subpial zone, it assumes a plexiform appearance (Fig. 2.1a, 22 days, c, d). At this stage, some subpial neurons start to develop specific morphological features. Some neurons, sandwiched among the fibers, assume a horizontal morphology and tend to occupy the subpial upper region, while others assume a stellate morphology and tend to occupy its deeper region (Fig. 2.1c). The early horizontal neurons are recognized as embryonic Cajal–Retzius cells while the deep stellate ones are recognized as pyramidal-like neurons with ascending apical and basal dendrites (Figs. 2.1c, d and 2.2a, b). The embryonic Cajal–Retzius cells are already characterized by horizontal dendrites and axonic terminals distributed throughout the subpial zone upper region (Fig. 2.2). At this time, the deep pyramidal-like cells become the larger neurons of the developing neocortex. These neurons axon branches through the subpial lower region, has ascending collaterals that reach the upper zone, and becomes the source of the early corticofugal fibers leaving the developing cortex (Fig. 2.2a, b). The subpial zone thickness has increased considerably and its fibrillar and neuronal elements have expanded horizontally throughout the cortex (Fig. 2.2a, b). At this age, a thin band of internal white matter composed of both corticipetal and corticofugal fibers start to be recognized through the neocortex proximal or ventral region. In Golgi preparations, some of its fibers have terminal growth-cones advancing in opposite directions suggesting either arriving and/or departing fibers (Fig. 2.2a, b).

At this developmental stage, the mammalian neocortex is considered to be already an early functional system. Because of its appearance and neuronal composition, this early cytoarchitectural and functional organization was originally named the "PP Layer" (Marín-Padilla 1971). It was later renamed the "Prelate." In mammals, the PP developmental stage is a transient organization recognized in 11-day-old mouse (Fig. 2.1d), 11-day-old hamster (F. 2A), 22-day-old cat (Figs. 2.1c and 2.2b), and in 50-day-old human embryos (Chapter 3).

The PP developmental stage precedes the appearance of the cortex PCP and will regulate its ascending and stratified organization, as well as the ascending placement of its pyramidal neurons from lower (older) to upper (younger) cortical strata.

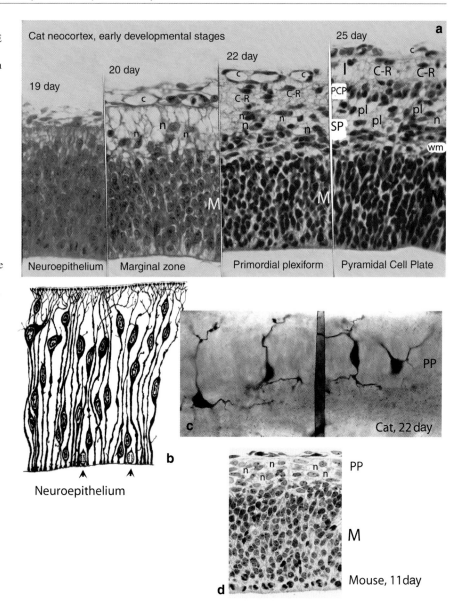

Fig. 2.1 Montage of photomicrographs from H&E (**a**, **d**) and rapid Golgi preparations (**c**) and a camera lucida drawing (**b**) showing various aspects of the of the mammalian cerebral cortex early developmental (embryonic) stages. (**a**) The progression of the cat neocortex embryonic development is illustrated through its neuroepithelium, marginal zone (MZ), primordial plexiform(PP), and pyramidal cell plate (PCP) developmental stages, respectively. (**b**) Camera lucida drawing illustrating the neuroepithelium cellular composition and the terminal filaments with endfeet that construct the cerebrum limiting pial surface and basal lamina. (**c**) Golgi preparations illustrating the pyramidal-like and stellate morphology of some early neurons of the cat cerebral cortex. (**d**) H&E preparation illustrating the hamster cerebral cortex early primordial plexiform (PP) embryonic stage with scattered neurons throughout the subpial zone, prominent matrix (M) zone, and numerous mitotic cells throughout the ependymal surface

Pyramidal Cell Plate (PCP) Stage. Migrating neuroblasts, of ependymal origin, start to arrive at the developing neocortex and begin to accumulate precisely within the PP forming a compact cellular plate that eventually extends throughout the entire cortex. This cellular plate, which represents a distinguishing mammalian feature, is renamed, in this monograph, the PCP (Fig. 2.3a). The new PCP terminology is preferable to the current CP, because during its formation it is essentially composed of pyramidal neurons (Chapter 3). Its appearance also follows a ventral (proximal) to dorsal (distal) gradient, which is recognized in sections of the entire embryo brain (Figs. 11.3a and 11.9). The PCP represents a mammalian innovation and its pyramidal cells are the essential and most distinguishing neuronal type of the cortex. Its appearance divides the original PP neuronal and fibrillar components into elements that remain above it and those that remain below it (Fig. 2.3a, b). Thus establishing simultaneously the cortex first lamina (I) above it and the subplate (SP) zone below it (Fig. 2.3a, b). All PCP pyramidal neurons originate intracortically from mitotic cells within the ependymal epithelium. The pyramidal cell original precursors, using radial glia filaments as guides and attracted by

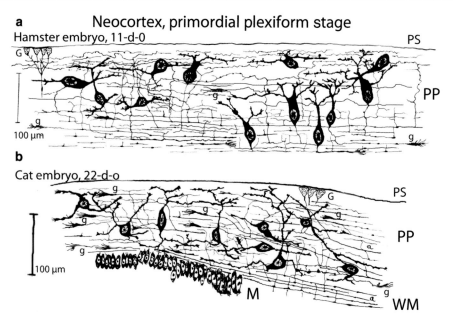

Fig. 2.2 Montage of camera lucida drawings, from Golgi preparations, of the hamster (**a**) and the cat (**b**) embryonic cerebral cortex showing the fibrillar and neuronal composition of their primordial plexiform developmental stage. Both neurons and fibers are scattered throughout the entire thickness of the subpial zone without specific configuration. (**a**) The neurons throughout the subpial zone of an 11-day-old hamster embryo cerebral cortex are still morphologically undermined and variable and scattered among numerous horizontal fibers. (**b**) The neuronal and fibrillar composition and organization of a 22-day-old cat embryo cerebral cortex are also still undifferentiated and without specific locations. In the cortex of both hamsters and cat embryos, all neurons and fibers are intermingled and scattered throughout the subpial zone lacking distinct location and/or specific stratification. In both embryos, some of the horizontal fibers have terminal growth cones advancing in opposite direction suggesting the presence of both corticipetal and corticofugal fibers. At this age, an incipient internal white matter band starts to be recognized, especially, in proximal regions of the developing cerebral cortex of both embryos. All components of the mammalian neocortex during its early developmental stage are concentrated between the pial surface and the cellular matrix (M) zone. Scales = 100 μm

reelin from Cajal–Retzius cells, ascend up to the PP zone and accumulate progressively within it (Rakic 1972, 1988; Marín-Padilla 1992). The newly arrived neuroblasts after losing their glial attachment are transformed into mammalian pyramidal neurons with an apical dendrite that branches within the first lamina forming a bouquet, establish functional contacts with Cajal–Retzius cells, and develop a rudimentary descending axonic process (Fig. 2.3); see also Chapters 3–5.

At this stage, the following strata are recognized in the mammalian cortex: (a) a first lamina with Cajal–Retzius cells with horizontal dendrites and axonic processes, the primordial corticipetal fibers horizontal axon terminals, the apical dendritic bouquets of newly arrived pyramidal neurons, the axon terminals of SP Martinotti cells, and the apical dendritic bouquets of SP pyramidal-like neurons (Fig. 2.3a, b); (b) an intermediate PCP stratum composed of immature pyramidal neurons attached to the first lamina by their apical dendrites (Fig. 2.3a, b); (c) a deep SP zone stratum with large pyramidal-like neurons with ascending apical dendrites that reach and branch within the first lamina, with several basal dendrites and an axon that after giving off several collaterals penetrates into the internal white matter and becomes a corticofugal fiber; (d) a thin internal white matter composed of both corticipetal and corticofugal fibers (Fig. 2.3a, b); (e) a thick matrix zone composed of closely packed cells; and (f) an ependymal epithelium with numerous mitotic cells and the attachment of numerous radial glial cells. The SP pyramidal-like neurons axon has ascending collaterals that reach the first lamina and several horizontal collaterals distributed through the zone (Fig. 2.3b). The SP zone has also Martinotti neurons with basal dendrites and an ascending axon that reaches and branches within first lamina (Fig. 2.3b).

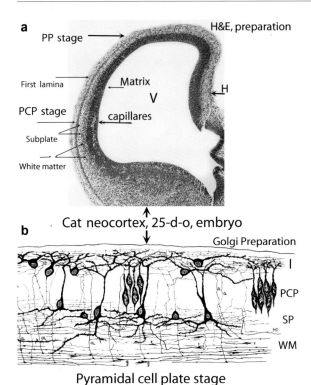

Fig. 2.3 Composite figure including an H&E preparation (a) of the cerebral vesicle (hemisphere) of the brain of a 25-day-old cat embryo, and a camera lucida drawing (b), from rapid Golgi preparation of the neocortex, of 25-day-old cat embryos, showing its evolving PP (PP) and PCP (PCP) development stages and their cytoarchitectural organizations and compositions. (a) The section represents a coronal cut of an entire cerebral hemisphere of a cat embryo to demonstrate the proximal (ventral) ongoing developmental gradient of the PCP formation while the neocortex distal (dorsal) region still is at the PP developmental stage. Also illustrated are the ongoing and simultaneous establishments of the first lamina and subplate (SP) zone as the PCP is being formed, above and below it, respectively. The hippocampus (H) anlage, the early intrinsic microvascularization of the matrix, paraventricular region and striatum, and the early internal white matter are also recognized. (b) The camera lucida drawing demonstrates the cytoarchitectural organization and composition and the four basic strata of the cat embryo neocortex at this age. They include: the first lamina (I) with a variety of neurons, horizontal fiber terminals, the terminal dendritic bouquets of SP pyramidal-like and of newly arrived pyramidal neurons, the PCP with the bodies of newly arrived pyramidal neurons; the SP zone with large pyramidal-like and Martinotti neurons; and a band of internal white matter (WM). The PCP is crossed by ascending primordial corticipetal fibers that reach and branch within the first lamina and by the ascending axons of Martinotti cells. Some white matter fiber terminals have growth cones advancing in opposite directions representing corticipetal and corticofugal fibers

Some of the white matter fibers depict growth-cones advancing in opposite directions, suggesting the presence of both corticipetal and corticofugal fibers (Fig. 2.3b). At this age, the cat motor cortex PCP consists of a compact cellular plate (3–5 cells thick) composed of pyramidal neurons with smooth (spineless) apical dendrites that branch into the first lamina with short descending axons. The PCP advancing formation, from proximal (ventral) to distal (dorsal) cortical regions, delineates the simultaneous and sequential establishment of the neocortex first lamina and SP zone (Fig. 2.3a). At this age, the cat embryonic cortex distal region still is at its PP developmental stage (Fig. 2.3a). At this age, the matrix, the proximal paraventricular zones and the striatum has started to develop their intrinsic microvascular system (Fig. 2.3a). Also, at this age, most the embryonic cortex is still avascular but covered by a rich pial capillary anastomotic plexus that extends throughout its entire surface (Fig. 2.3a).

In the original descriptions, the cytoarchitecture composition of the cat cortex PP and early PCP were considered to share features with the amphibian and reptilian primitive cortices (Marín-Padilla 1971, 1978, 1992, 1998). As in reptiles, the embryonic mammalian cortex basic functional elements include Cajal–Retzius neurons on its outer zone, pyramidal-like projective neurons and Martinotti cells on its inner zone, and an internal white matter with both corticipetal and corticofugal fibers. At this age, the few pyramidal neurons of the PCP are considered to be undifferentiated and functionless with descending axons that barely reach the internal white matter.

The fact that the early cytoarchitectural organization of the embryonic mammalian neocortex should resemble that of phylogenetically older species is not surprising since it reflects a continuous process that interconnect the evolution of these three vertebrates (Marín-Padilla 1971, 1972, 1978, 1992, 1998). This developmental conception led to hypothesize the dual origin and composition of the mammalian cerebral cortex (Marín-Padilla 1971, 1978). Consequently, the first lamina and SP zone of the mammalian neocortex represent elements of a primordial cortical organization that shared features with the primitive cortex of amphibian and reptiles. To this primordial cortical organization, a PCP, which represents a mammalian innovation, is progressively incorporated. The newly formed PCP up to around the 15-week-of gestation is composed essentially of undifferentiated pyramidal neurons anchored to the first lamina by their apical dendrites and smooth somata with short descending axons. Eventually these pyramidal neurons become the major component of the cortical gray matter (Chapters 3 and 4). Consequently, the term "neocortex" often used to describe the mammalian cerebral cortex seems most appropriate.

These four early embryonic stages evolve rapidly (in days) and throughout a still avascular neocortex. As a neuroectodermal tissue, the early cortex is deprived of an intrinsic microvascular system and, eventually, will have to develop one. The embryonic cerebral cortex is surrounded by well-vascularized meninges (Chapter 7). The pial anastomotic capillary plexus evolves from meningeal arachnoidal vessels and progressively covers the entire cortex surface, the primordial hippocampus, and supplies the choroid plexus vasculature. This rich anastomotic pial capillary plexus covers the neocortex surface through its early (NE, MZ, PP and early PCP) embryonic stages. These pial capillaries are separated from the cortical elements, only by the pial EGLM and its basal lamina. Undoubtedly, the close proximity of these capillaries to the cortex early neuronal and fibrillar elements allows sufficient oxygen diffusion to guarantee their early evolution. Eventually, capillary sprouts from this plexus will perforate through the neocortex basal lamina and the EGLM and penetrate into the nervous tissue, progressively establishing the various compartments of the cerebral cortex extrinsic and intrinsic microvascular systems (Chapter 7).

References

Marín-Padilla M (1971) Early prenatal ontogenesis of the cerebral cortex (neocortex) of the cat (Felis domestica). A Golgi study. Part I. The primordial neocortical organization. Z Anat Entzwickl-Gesch 134:117–145

Marín-Padilla M (1972) Prenatal ontogenesis of the principal neurons of the neocortex of the cat (Felis domestica). A Golgi study. Part II. Developmental differences and their significance. Z. Anat Entzwickl-Gesch 136:125–142

Marín-Padilla M (1978) Dual origin of the mammalian neocortex and evolution of the cortical plate. Anat Embryol 152: 109–126

Marín-Padilla M (1983) Structural organization of the human cerebral cortex prior to the appearance of the cortical plate. Anat Embryol 168:21–40

Marín-Padilla M (1992) Ontogenesis of the pyramidal cell of the mammalian neocortex and developmental cytoarchitectonic: a unifying theory. J Comp Neurol 321:223–240

Marín-Padilla M (1998) Cajal-Retzius cell and the development o the neocortex. Trends Neurosci (TINS) 21:64–71

Marín-Padilla M, Marín-Padilla T (1982) Origin, prenatal development and structural organization of layer I of the human cerebral (motor) cortex. Anat Embryol 164:161–206

Rakic P (1972) Mode of cell migration to the superficial layers of the fetal monkey neocortex. J Comp Neurol 145:61–84

Rakic P (1988) Intrinsic and extrinsic determinants of neocortical parcellation: a radial unit model. In: Rakic P, Singer W (eds) Neurobiology of the neocortex. Wiley, New York, pp 92–114

Zecevic N, Milosevic A, Rakic S, Marín-Padilla M (1999) Early development and composition of the primordial plexiform layer: an immunohistochemical study. J Comp Neurol 412: 241–254

Human Motor Cortex: Development and Cytoarchitecture

The prenatal development of the human brain is characterized by its extraordinary growth, surface expansion, and increase in the number of circumvolutions, which further enhance its surface area (Fig. 3.1). Its surface and underlying gray matter – where most neurons reside – expand from about 20 cm^2 at 12-week gestation to 716 cm^2 by birth time; and, its weight increases from 2 g at 10-week gestation to 410 g by birth time and its weight will triplicate by adult time (Blinkov and Glezer 1968). The large number of pyramidal neurons incorporated into the gray matter and their anatomical and functional expansions also contribute to its surface expansion and increasing circumvolutions. In addition, the number of pyramidal neurons incorporated into the gray matter increases significantly from lower (older) to upper (younger) strata further contributing to the progressive fanning of its surface and the formation of circumvolutions (see Fig. 4.10). Eventually, the large number of neurons incorporated into the developing gray matter will exceed the cranial cavity available space, forcing the folding of the brain upper strata where most neurons reside.

The prenatal development of the human brain goes through an early embryonic and a late fetal period. The embryonic period expands from the fifth to the eighth week of gestation and fetal period from the eighth week of gestation to birth time (Larsen 1993; O'Rahilly and Müller 1994). The embryonic period begins with the formation of the cerebral vesicles followed by the establishment throughout its subpial zone, in a proximal (ventral) to distal (dorsal) gradient, of three successive developmental stages, which in order of appearance are: the neuroectodermal, marginal zone (MZ), and primordial plexiform (PP) stages (Chapter 2). The embryonic period of the brain concludes with the appearance of the pyramidal cell plate (PCP)(Chapter 2). The main events that occur during the human embryonic period are tabulated and timed in Fig. 3.2. The fetal period is essentially characterized by the ascending and stratified development and maturation of the gray matter pyramidal neurons (Marín-Padilla 1970a, 1972b, 1978, 1990a, 1992). The developmental incorporation into the gray matter of inhibitory interneurons and both the intrinsic microvascular and the neuroglial systems also follow a concomitant, ascending, and stratified maturation that runs parallel that of the pyramidal neurons (Marín-Padilla 1985b, 1995).

Rapid Golgi preparations of postmortem unspoiled brain tissue, from 27 selected cases of different gestational ages were obtained and prepared during 36 consecutive years. The number of rapid Golgi preparation obtained from these cases and studied is over 4,500 (Chapter 12). The cases studied include embryos and fetuses, which died due to a variety of clinical problems and were postmortem evaluated. Their gestational ages, include: the 6th-, 7th-, 11th-, 15th-(2), 16th-(2), 17th-, 18th-, 20th-(3), 22nd-, 24th-, 26th- (2), 29th-, 30th- (2), 32nd-, 35th- (3), and 40th-week gestation (4 newborn). The recorded gestational ages are close approximations and the numbers in parenthesis refer to the number of cases studied with a similar gestational age. Only the prenatal development of the primary motor cortex of these brains were sequentially studied. The results of these life-long developmental Golgi studies are reassembled, reorganized by age, described, and amply illustrated in this chapter.

All cases studied were from postmortem brain specimen from the Department of Pathology of Dartmouth-Hitchcock Medical Center. As Director of its Pediatric Autopsy Service (1962–1999), I was familiar with most pediatric cases, have direct access to all pediatric autopsies and the responsibility for

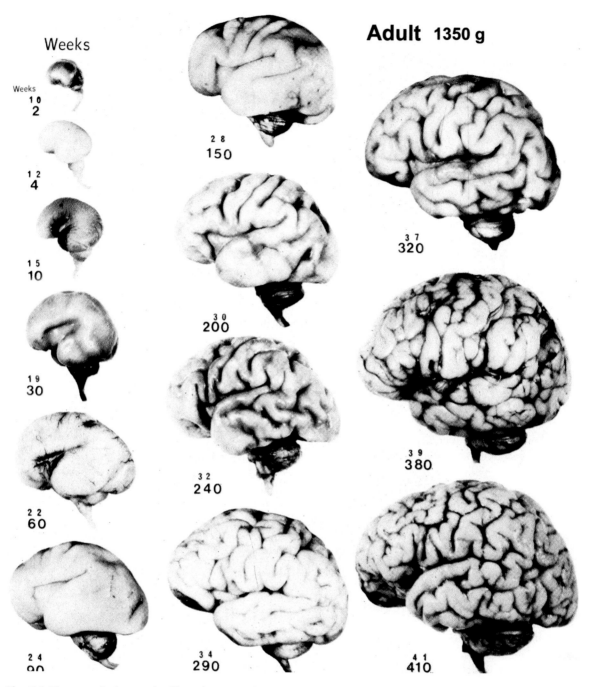

Fig. 3.1 Montage of photographs illustrating, at scale, the human brain prenatal development, including: age, size, weight, morphologic features, and evolving formation of the circumvolutions and gyral patterns. It is composed of individual photos of formalin-fixed human brains from the 10th to the 40th (newborn) week of gestation. The average weight of an adult human brain is also included for comparison. It is fundamental to correlate the brain size and the degree of cytoarchitectural organization and functional maturation of its cerebral cortex gray matter. (Modified from Larroche, 1977)

their pathological postmortem evaluations. Major clinical causes for the infants' demise include: prematurity, respiratory difficulties, labor complications, and/or perinatal infections. The brains of all cases included in this monograph were considered to be pathologically normal and/or unaffected. All the

Fig. 3.2 Schematic figure tabulating the main events that occur during the embryonic period of human development, including the evolutions of the nervous system and cerebral cortex. The embryo length and gestational age are correlated with the various events. (Modified from Marín-Padilla 1983)

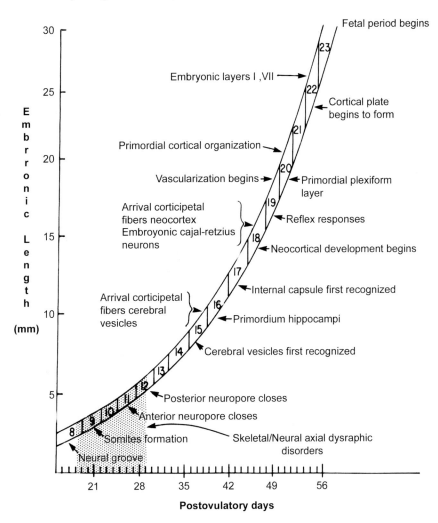

necessary legal and institutional protocols, as well as the parental consents, were fulfilled in all case (Chapter 12).

In addition to the rapid Golgi procedure, other staining procedures were also used in some cases, including: routine staining procedures (hematoxylin & eosin, Nills, Luxo-fast blue and Bodian silver), some immunohistochemical ones (glial fibrillary acidic protein, neurofilament for SMI-32 nonphosphorylated proteins, parvalbumin, and synaptophysin), and the Golgi-Cox variant for formalin-fixed tissue. The combined usage of these procedures permitted to observe some cytoarchitectural details, which was impossible to detect using single techniques (Chapter 12).

Besides Cajal's original Golgi studies of newborn infants' cerebral cortex motor, visual, acoustic, and olfactory regions (Cajal 1899a, b, Cajal 1900a, b) and mine (Marín-padilla 1967, 1969 1968, 1969a, b, 1970b, 1972a, 1974a, b, 1976, 1985a, b, 1987a, b, 1990b, 1996, 1997, 1999, 2000, Marín-padilla and Marín-padilla 1980, Marín-padilla et al. 2002), which span from 1967 to 2002, no other systematic studies of the developing human brain have been reported in the literature, although some aspects of the developing human brain cytoarchitectural organization, using this classic procedure, have been occasionally reported (Retzius 1893, 1894; Kölliker 1896; Poljakow 1961; Mrzljak et al. 1988).

The description of the developing cytoarchitectural organization and composition of the human brain motor cortex have been distributed into seven separate developmental stages of increasing age. Each stage records the findings of cases with roughly a similar gestational age. The gestational age of each of the case described will also be recorded. The descriptions in this chapter (and in Chapter 4) emphasize the developmental history of the human motor cortex pyramidal neurons. The developmental histories of other motor cortex elements are described in Chapters 5–8.

3.1 Seven-Week-Old Stage

Essentially, the human cerebral cortex cytoarchitectural organization starts at this developmental stage. Its basic composition and cytoarchitectural organization have been explored in two embryos, aged 43 and 50 days, respectively (Fig. 3.3a, b).

The first embryo (15 mm crown–rump length) is still quite immature with terminal paddles in all extremities, a short tail, and still lacking distinguishing human features (Fig. 3.3a). This embryo was recovered from an

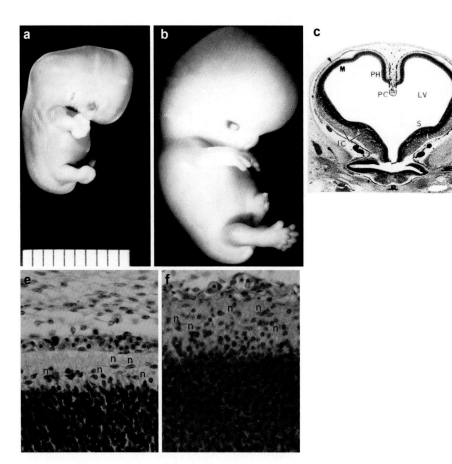

Fig. 3.3 Composite figure with various photographs showing the size and anatomic features of two human embryos, aged 43 days (**a**) and 50 days (**b**), the development of their respective cerebral cortices (**c**, **d**) and photomicrographs of their corresponding developmental stages (**e**, **f**). While the 43-day-old embryo (**a**) is still quite immature and undifferentiated, the 50-day-old one has already developed unmistakable human features with a large head and prominent forehead. In only 7 days the human embryo has duplicated its size. The first embryo cerebral cortex (**e**) is at the marginal zone (MZ) developmental stage characterized by a distinct subpial zone with a scattered horizontal and poorly differentiated neurons and by a compact matrix zone. The second embryo (**f**) cerebral cortex is at the primordial plexiform (PP) stage of development and is characterized by a significant increase in the number of both neurons and of horizontal fibers throughout its subpial zone. The cerebral cortex, in both embryos (**e**, **f**), still is avascular but covered by rich pial capillary anastomotic plexus, with nucleated red. Key: LV, lateral ventricle; PH and H, primordial hippocampus and hippocampus; PC, primordial choroid plexus; S, striatum; IC internal capsule. Scale in mm. (Modified from Marín-Padilla 1983)

unruptured tubal pregnancy removed surgically. His brain is also quite immature, its laterals and third ventricles are large, the cortical mantle is very thin and the anlage of the hippocampus, choroid plexus, and pyriform lobe are also recognized (Fig. 3.3c). A rich pial anastomotic capillary plexus, with nucleated red cells, covers the entire still avascular cerebral cortex (Fig. 3.3c, e). The primordial striatum has already started to develop its intrinsic microvasculature. A few advancing fibers from the newly established internal capsule have penetrated into the cerebral vesicle and some of them have reached its subpial region (Fig. 3.3c). These primordial corticipetal fibers enter into the developing cortex through its subpial region following a ventral (proximal) to dorsal (distal) progression. Serial sections of the embryo's entire head corroborate the subpial progressions of these primordial corticipetal fibers. At this age, the cerebral cortex of the human embryo is at the MZ developmental stage (Chapter 2).

In addition, in the cortex proximal (ventral) regions, a few scattered undifferentiated neurons are recognized through the subpial zone (Fig. 3.3e). Some of these early neurons, sandwiched among the horizontal primordial corticipetal fibers, have assumed a horizontal morphology compatible with that of embryonic Cajal–Retzius (C-R) cells (Fig. 3.3e). There are also others morphologically less defined neurons scattered through the subpial zone above the compact matrix zone (Fig. 3.3e). The matrix zone is composed of numerous and closely packed cells with nuclei at all levels. The ependymal epithelium has numerous mitotic cells (Fig. 3.3e). This human embryo cortical development is comparable to that of 20-day-old cat embryos (Chapter 2).

The second human embryo (22 mm crown–rump length) was recovered from an intact pregnant uterus of a hysterectomy surgical case (Fig. 3.3b). The pregnant uterus was carefully dissected under water to preserve intact and undisturbed the embryo amniotic cavity. Unfortunately, this little astronaut journey, inside its protective and transparent amniotic capsule, was interrupted 50 days after his departure. His contribution to a better understanding of the human brain early cytoarchitectural organization and composition is herein duly acknowledged.

This male embryo was 50 days old and his cerebral cortex was at the PP developmental stage (Chapter 2). Comparing both embryos, in only a few days, the brain of the second has grown considerably (compare Fig. 3.3a, b). His size has nearly doubled, the retinas are already pigmented, fingers and toes are already formed, the tail size is significantly reduced, and his prominent large head is already unmistakably human (Fig. 3.3b). No other mammalian embryo has, at this early age, such a prominent and large head. It can be stated that, by 50 days of age, the human embryo is distinctly human.

The head enlargement and specially its frontal prominence are caused by the expansion of both the cerebral cortical mantle and ventricles. The enlargements of the hippocampus, dorsal and ventral striatum, and choroid plexus have also contributed to his large head (Fig. 3.3d). The number of fibers passing through the internal capsule and reaching the cortex subpial zone has also increased considerably as well as the number of recognizable scattered new neurons throughout it (Fig. 3.3d, f). This embryo subpial zone, compared to that of the previous one, has tripled in its thickness as well as in the number of both new neurons and primordial corticipetal fibers (Marín-Padilla 1983).

Parasagittal brain sections, at the internal capsule entrance level, demonstrate the large number of corticipetal fibers entering into the cerebral vesicle (Fig. 3.4, inset). Using Bodian silver stain preparations, the progression of the internal capsule fibers toward the cortex subpial region has been recorded at four different levels (Fig. 3.4a–d). Most of the entering fibers (Fig. 3.4a) are distributed through ventral and dorsal striatum (Fig. 3.4b); some reached the pyriform lobe (Fig. 3.4c) and a few of them have reached and penetrated into the cortex through its subpial zone (Fig. 3.4d). Some subpial zone neurons have already assumed a horizontal morphology and tend to occupy its outer region, while others of undetermined and/or stellate morphology tend to occupy its lower region, immediately above the matrix (Fig. 3.4e–h). The superficial horizontal neurons are considered to represent embryonic C-R cells and the deepest ones the embryonic subplate (SP) zone neurons (Chapter 2). Throughout the subpial zone there are silver stained horizontal fibers, believed to represent primordial corticipetal and some early corticofugal fibers from its neurons, already leaving the developing cortex (Fig. 3.4e–h). The lack of rapid Golgi preparations makes further identification of these neurons and fibers impossible. A rich pial anastomotic capillary plexus covers the still avascular cerebral cortex (Fig. 3.4e–h).

At this embryonic age, the cytoarchitectural composition and organization of the human cerebral cortex corresponds to that of a PP stage. This primordial cortical organization is considered to be already a functional active system. This early functional system is

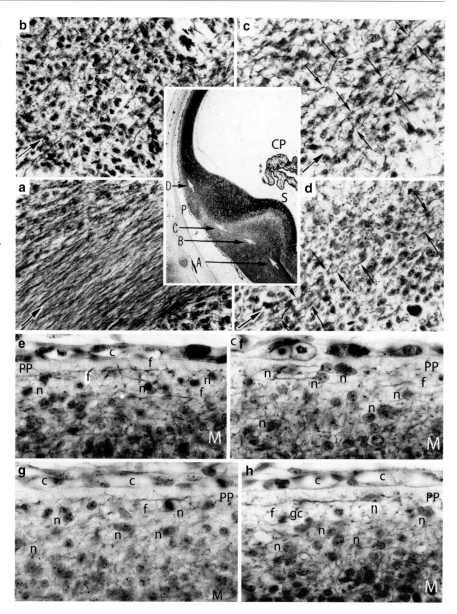

Fig. 3.4 Montage of photomicrographs (from Bodian silver preparations) of the cerebral cortex of a 50-day-old human embryo showing the progression of corticipetal fibers from the internal capsule into the cerebral cortex subpial zone (**a–d** and inset) as well as their presence throughout this zone (**e–h**). Most corticipetal fibers passing through the internal capsule (**a** and inset) are distributed through the ventral and dorsal striatum (**b**), some have reached the pyriform lobe (**c**) and a few of them have entered into the cerebral cortex through the subpial zone (**d**). The large arrows point toward the direction of the advancing fibers and the short one to the fibers themselves. e, f, g, and h, illustrate the presence of both horizontal fibers (f) and scattered neurons (n) throughout the cortex subpial zone. Some neurons (**f, h**) have already assumed a horizontal morphology compatible with Cajal–Retzius cells and tend to occupy the subpial zone upper regions. While other less differentiated ones (**e, f**) tend to occupy the zone lower region. The cerebral cortex of this 50-day-old embryo is at the PP stage of development

composed of primordial corticipetal fibers (afferent elements) that target the dendrites of both the embryonic C-R cells and the deep pyramidal-like neurons (receptor elements) and of early corticofugal fibers (efferent elements) from these neurons axonic terminals. This early developmental stage has been confirmed in human embryos (Larroche 1981).

This developmental stage of the human brain is comparable to that of cats and rats embryos (Marín-Padilla 1971; Luskin and Shatz 1985; Bayer and Altman 1991). At this age, the developing human cerebral cortex is comparable to that of 22-day-old cat and 11-day-old mouse and hamster embryos (Chapter 2).

3.2 Eleven-Week-Old Stage

This is the early stage of the developing human cerebral cortex that has been studied with the Golgi procedure. This 11-week-old fetus was recovered from a spontaneous abortion and its preservation was not optimal due to slight postmortem tissue deterioration. Although the Golgi preparations are of poor quality, it has been possible to establish the human motor cortex basic cytoarchitectural organization and composition at this age (Fig. 3.5).

At this age, the human motor cortex, through an intermediate region, is about 350 μm thick. The following

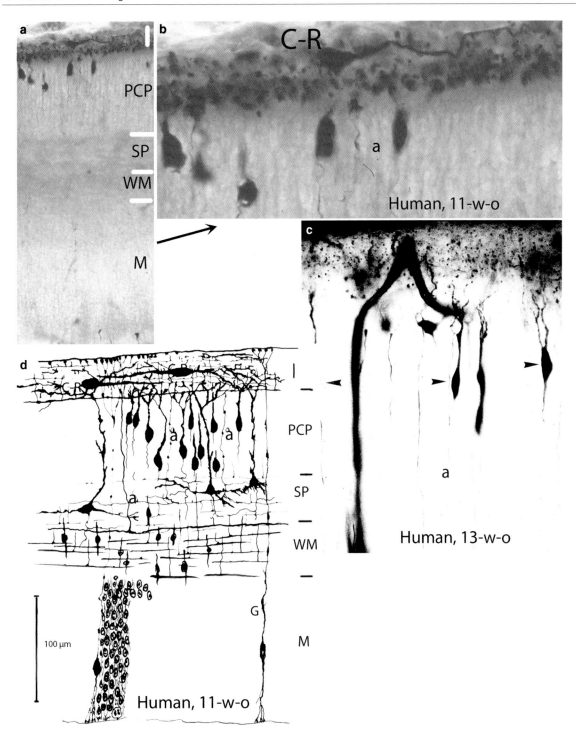

Fig. 3.5 Composite figure of photomicrographs from Golgi-Cox preparations of the cerebral cortex of an 11-week-old (**a, b**) and a 13-week-old (**c**) human fetuses and a camera lucida drawing (**d**), from the 11-week-old fetus showing the basic cytoarchitectural organization and composition of the human cerebral cortex at this age. The camera lucida drawing (**d**) illustrates, at scale and more clearly, the motor cortex basic cytoarchitecture organization, including: the first lamina (I) Cajal–Retzius cells and horizontal fiber, the thick pyramidal cell plate (PCP) composed of immature pyramidal neurons attached to first lamina, the ascending corticipetal fibers that reach and branch within the first lamina, the subplate (SP) zone pyramidal-like neurons with apical dendrites that reach first lamina and descending axons that reached the white matter (WM) and become the first corticofugal fibers leaving the cortex, a narrow WM zone and a thick matrix zone (M). Scale in 100 μm

features are recognized in the motor cortex at this age: a first lamina (40 μm thick) with C-R cells and horizontal fibers (Fig. 3.5 a, b, d); an early PCP (75 μm thick) composed of 6–8 levels of closely packed pyramidal neurons; a subpial (SP) zone (45 μm thick) with large pyramidal-like neurons (Fig. 3.5d); a thin band (50 μm thick) of internal white matter (WM) with both corticipetal and corticofugal fibers that extend throughout the entire cortex (Fig. 3.5d); a thick matrix zone composed of closely packed cells; and, an ependymal epithelium with numerous mitotic cells. The motor cortex is crossed by numerous radial glial filaments with terminal endfeet incorporated into the pial external glial limiting membrane (EGLM) and by ascending primordial corticipetal fibers from the WM that reach the first lamina and become long horizontal terminals (Fig. 3.5a, b, d).

The PCP (gray matter) is composed of tightly packed immature pyramidal cells with smooth (spineless) apical dendrites that reach the first lamina and branch forming a dendritic bouquet with spine-like excrescences. These pyramidal neurons have smooth somata and short descending axons (Fig. 3.5d). The SP zone large pyramidal-like neurons are the largest and most developed neurons of the cerebral cortex at this age. These neurons have an apical dendrite that reaches and branches within the first lamina and several basal dendrites distributed through the SP zone. They have a descending axon that penetrates into the WM and become a corticofugal fiber (Fig. 3.5d). The axon has also a recurrent ascending collateral that reaches the first lamina and several collaterals distributed through the SP zone.

At this age, the C-R cells and the SP zone pyramidal-like neurons are anatomically and functionally interconnected; both received inputs from primordial corticipetal fibers. Together these two neurons constitute the motor cortex primordial functional system. At this age, the SP zone pyramidal-like neurons are the only source of corticofugal fibers leaving the developing neocortex. The few pyramidal neurons of the developing gray matter are still immature with spineless apical dendrites, smooth somata, and short descending axons that are barely reaching the WM. Consequently, they are not considered to be components and/or to participate in the motor cortex primordial functional system. This fetus motor cortex is already vascularized by perforating capillaries that reached the matrix zone and have developed interconnecting anastomotic capillary plexuses, at various cortical levels. The first lamina, although perforated by pial capillaries, have not yet developed its intrinsic microvascular system.

Golgi preparations of the developing neocortex of a 13-week-old human fetus (reported in the literature) describe similar cytoarchitectural features and basic cortical organization (Mrzljak et al. 1988). The cortex of this fetus also has immature pyramidal neurons with spineless apical dendrites anchored to first lamina, smooth somata, and short descending axons (Fig. 3.5c). No further morphological details of the brain of this fetus are recorded.

3.3 Fifteen-Week-Old Stage

This developmental stage is represented by six fetuses; aged 15- (2), 16- (2), 17- and 18-week old, respectively, and by good quality rapid Golgi preparations. Because of the large number (1,360) of preparations available and the good quality of some of them, this developmental stage of the human motor cortex is well documented and has been particularly well studied (Figs. 3.6–3.9).

At this age, the human brain still has a primitive appearance with large frontal and temporal poles separated by a wide lateral fissure (Fig 3.1). It has a smooth surface and only weighs about 10 g. The cerebral cortex thickness ranges from 1,200 μm, at proximal (ventral) regions, to 600 μm at distal (dorsal) ones (Fig. 3.6a). Hematoxylin and eosin (H&E) preparations through its mid-section demonstrate the following strata: a first lamina (50 μm thick) with C-R cells and horizontal fibers; a gray matter (300 μm thick) composed of tightly packed immature pyramidal neurons with spineless apical dendrites anchored to first lamina, smooth somata and short descending axons; a SP zone (300 μm thick) with large pyramidal-like neurons and numerous horizontal and vertical fibers terminals; a band of WM composed of both corticipetal and corticofugal fibers extending throughout the entire neocortex; a thick germinal matrix (paraventricular) zone composed of closely packed cells and large thin-walled blood vessels; and an

3.3 Fifteen-Week-Old Stage

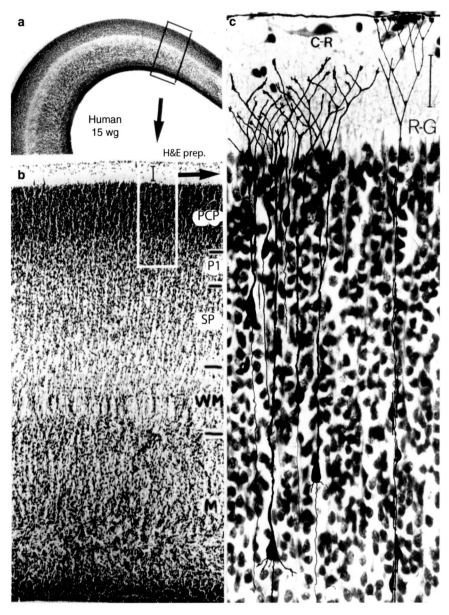

Fig. 3.6 Composite figure of photomicrographs from H&E preparations, showing three different views (magnifications) of the cerebral cortex of a 15-week-old human fetus. (**a**) Illustrates a low power view of the infant cerebral cortex showing its basic cytoarchitectural organization and stratification, more advanced through ventral (proximal) than through dorsal (distal) regions. (**b**) Reproduces at a higher magnification the rectangular section outlined in (**a**), showing its apparently barren first lamina (I), the compact pyramidal cell plate (PCP) already about 100 cell thick with more loose cellular arrangement throughout its lower and older regions indicates the presence of horizontal fiber terminals and the establishment of the first pyramidal cell P1 functional stratum. The PCP lower zone, subpial zone (SP), white matter (WM) and the matrix zone show many vertical cellular aggregates separated by cell-free areas indicating the presence of unstained fibers from radial glia filaments, corticipetal fiber terminals, and descending pyramidal cell axons. (**c**) Reproduced at a higher magnification the rectangular area outlined at (**b**) showing the apparently empty compositions of the first lamina with a large Cajal–Retzius (C-R) cell, smaller glial cells, the terminal filaments of radial glial fibers with endfeet incorporated into the EGLM, and pyramidal neurons terminal dendritic bouquets. See also Fig. 3.8a and b

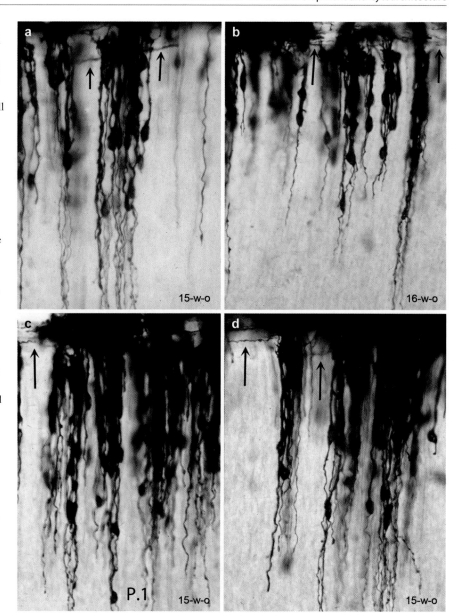

Fig. 3.7 Montage of photomicrographs from rapid Golgi preparations of the motor cortex of 15-week-old (**a**, **c**, **d**) and 16-week-old (**b**) human fetuses showing different views of their overall cytoarchitectural organization. At this age, the motor cortex PCP pyramidal neurons are still immature with smooth spineless apical dendrites that reach first lamina, smooth somata without basal dendrites and short descending axons that are starting to reach the white matter zone. The axons from the upper, smaller, and younger pyramidal neurons have not yet reached the PCP lower limit (**a**, **b**, **d**). At this age the PCP thickness is about 300 μm. Although silver deposits obscure the first lamina cytoarchitectural organization, it is possible, in some sections (*arrows*), to recognize the thick horizontal axonic terminals of Cajal–Retzius neurons coursing through its lower zone and targeting the terminal dendrites of pyramidal neurons; there are no other neuronal types in the human motor cortex gray matter at this age. The illustrations are self-explanatory

ependymal epithelium with numerous mitotic cells (Fig. 3.6a–c). The cortex is crossed, from ependymal to pial surface, by radial glial fibers with endfeet incorporated into cortex EGLM that manufacture its basal lamina material. Primordial corticipetal fibers ascend from the WM cross, unbranched, the gray matter, and terminate within the first lamina (Figs. 3.6b, c and 3.7a–d).

At this age, the motor cortex gray matter pyramidal cell plate (PCP) is about 100 cells thick and solely composed of immature and closely packed pyramidal cells with spineless apical dendrites anchored into the first lamina (Fig. 3.6b, c). The PCP formation is nearly complete and no additional pyramidal neurons are incorporated into developing gray matter after the 17th–18th–week of gestation. In H&E preparations, the gray matter tightly packed pyramidal cells are essentially arranged in columns separated by thin vertical cell-free zones occupied by the unstained filaments of the radial glia, ascending corticipetal fibers and pyramidal neurons descending axons (Fig. 3.6b, c). Throughout the cortex proximal regions, the lower and older gray matter pyramidal cells are more loosely arranged than

3.3 Fifteen-Week-Old Stage

Fig. 3.8 Composite figure showing: (**a**) a photographic montage of pyramidal neurons, from rapid Golgi preparations, of the motor cortex of 15-week-old human fetuses showing their size, location (depth), and their apical dendritic attachment to the first lamina; and, (**b**) a camera lucida drawing illustrating the basic cytoarchitectural organization and composition of the human motor cortex at this age. The size (length of apical dendrite) of pyramidal neurons within the PCP ranges from 30 μm for the superficial, younger, and smaller ones to 270 μm of for the deeper, older, and larger ones (**a** and **b**). At this age, while most pyramidal neurons are still immature, some of the deeper and older ones (through proximal regions) have started their ascending functional maturation by developing short basal dendrites and a few proximal apical dendritic spines and, consequently establishing the human motor cortex first pyramidal cell P1functional stratum. The PCP expansion (**b**) has displaced the subplate zone (SP) downward and, at this age, is located between 600 and 700 μm from the pial surface. Some subplate zone pyramidal-like and Martinotti neurons (**b**) have started to lose their functional contacts with first lamina to regress their processes and to assume the morphologic features of interstitial (polymorphous) neurons (Po). The SP zone neurons (**b**) are interspersed among the increasing number of fiber terminals throughout the region. At this age, horizontally migrating inhibitory neurons and/or glial precursors (**b**) are recognized through the PCP lower regions. The recognition of horizontally migrating neurons coincides with the functional maturation of motor cortex first pyramidal cell P1 stratum. The first lamina Cajal–Retzius cells (**b**) depict embryonic morphologic features characterized by the concentration of their dendritic and axonic processes, which precedes their eventual developmental lengthening

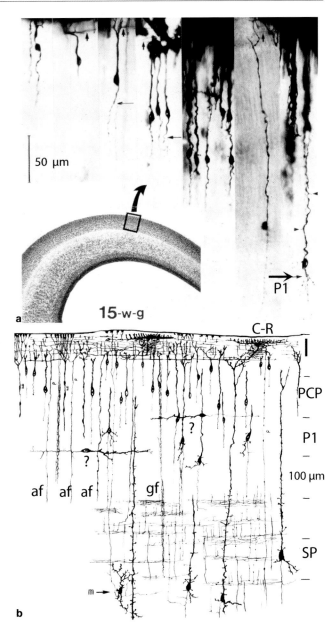

the upper ones, denoting the presence of unstained vertical and horizontal fibers among them (Fig. 3.6b). Both, the SP and the WM zones are also characterized by short vertical cellular bundles, separated by cells-free zones, probably representing ascending neuronal precursors around radial glia filaments (Fig. 3.6b). Throughout the SP and the WM zones, there are also numerous horizontal and vertical free-cell areas representing unstained corticipetal and corticofugal fiber terminals, the axonic collaterals of SP zone pyramidal like neurons and the descending axons of pyramidal neurons. These cell-free areas are more numerous and prominent through proximal (ventral) regions than in distal (dorsal) ones denoting differences in number of arriving corticipetal and departing corticofugal fibers passing through these two regions (Fig. 3.6a).

Rapid Golgi preparations confirm that the gray matter PCP is only composed of immature pyramidal neurons with spineless apical dendrites anchored in first lamina, smooth somata, and short descending axons (Figs. 3.7a–d and 3.8a, b). The size of the pyramidal neurons ranges from 30 μm, for the superficial

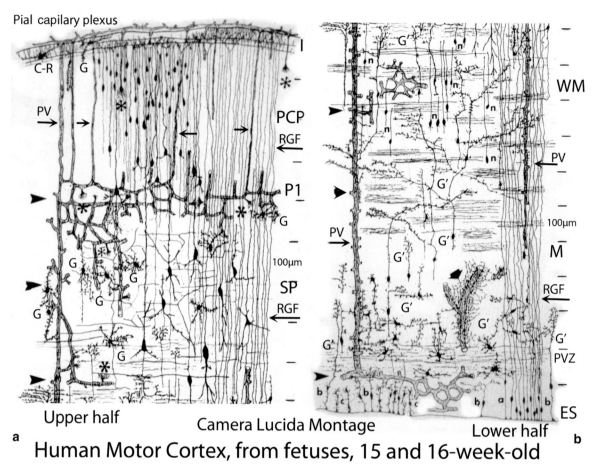

Fig. 3.9 Montage of a camera lucida drawings from rapid Golgi preparations of a 15-week-old human fetus, reproducing a section of the motor cortex, from pial to ependymal surfaces, measuring roughly 800 μm in width, 2,000 μm in height and 150 μm in thickness. The original large drawing has been separated into upper (**a**) and lower (**b**) portions to facilitate its view and understanding. The drawing illustrates the motor cortex overall cytoarchitectural organization as well as the morphology, distribution, and interrelationships of its basic neuronal, fibrillary, microvascular, and glial components. Only the PCP deep and older pyramidal neurons have developed short basal dendrite and some spines denoting the starting functional maturation and the establishment of the first pyramidal cell P1 stratum. Intrinsic microvascular anastomotic capillary plexus are already developed throughout different regions (*arrow heads*) of the motor cortex. The newly established one through PCP lower region coincides with the functional maturation of its lower and older pyramidal cells. The pyramidal neurons, above this stratum, remain immature and avascular. Fibrous astrocytes with vascular endfeet are recognized throughout the subplate and white matter (WM) zone as well as newly differentiated protoplasmic astrocytes through the PCP lower and older region. Ascending neuronal (n) and glial precursors with complex leading processes are found throughout the subplate and WM. Some glial precursors upon reaching the WM bend their leading processes in opposite directions possibly accompanying corticipetal as well as corticofugal fibers. Some of them may represent oligodendrocytes precursors. A prominent band of horizontal fibers, of unknown origin and destination, cross the paraventricular region (PVZ). The ependymal surface is characterized by the presence of numerous radial glial fibers (RGF) that ascend from ependymal to pial surface, and by numerous glial precursors, still attached to its surface, with complex leading processes. The remaining features of the drawing are self-explanatory

and youngest one, to 275 μm for the deepest and oldest ones, with intermediate sizes between them (Figs. 3.8a, b and 3.9). The length of the apical dendrites of the pyramidal neurons denotes their arrival time at first lamina and, hence, their age. The larger ones were the first and the shorter ones the last to reach the first lamina. Early arriving pyramidal neurons (and all subsequently arriving ones) will have to elongate anatomically their apical dendrite, without losing their functional anchorage to first lamina or somata location, to accommodate the arrival of new ones (Marín-Padilla 1992, 1998). The need to reach the first lamina and establish functional contacts with C-R cells will determine the age of the neuron, size as well as their inside-outside placement within the developing gray matter (Fig. 3.8a).

3.3 Fifteen-Week-Old Stage

At this age, the cortex gray matter (PCP) is composed of pyramidal neurons and no other neuronal types are recognized in it. Although in most rapid Golgi preparations silver precipitates obscure the first lamina cytoarchitectural organization, it is possible to recognize in some of them the presence of C-R cells thick horizontal axonic terminals that functionally target the apical dendritic bouquets of the pyramidal neurons throughout the neocortex (Fig. 3.7a–d, arrows).

At this age, the gray matter pyramidal neurons are still immature (Figs. 3.7a–d, 3.8a, 3.9). However, through proximal and mid-cortical regions, some of the deepest and oldest pyramidal neurons have started to develop short basal dendrites and a few proximal dendritic spines (Fig. 3.8a, b). The starting functional maturation of the deeper and older pyramidal neurons of the gray matter denotes that they have already established functional contacts with ascending corticipetal fiber terminals (Figs. 3.8a, b and 3.9). New corticipetal fibers, from the white matter, ascend and penetrate into the gray matter from lower to upper strata. These new corticipetal fibers are considered to be of thalamic origin and will functionally target the gray matter pyramidal neurons. These thalamic fibers branch through the gray matter lower strata and are distinct from the primordial corticipetal fibers that ascend unbranched through the gray matter to terminate into the first lamina. The arrival of these thalamic corticipetal fibers starts the ascending functional maturation of the gray matter pyramidal neurons. The starting functional maturation of the deepest and older pyramidal neurons of the gray matter (PCP) establishes the first functional pyramidal cell stratum in the human motor cortex, which should correspond to layer V of the current nomenclature. In the present monograph, we have renamed this first functional stratum of the motor cortex, the pyramidal cell P1 stratum (Figs. 3.8a, b and 3.9). It is important to emphasize that only the apical dendrite proximal segments of these neurons have started their functional maturation while their distal segments remain immature, spineless, and functionless. However, the terminal dendritic segments of all pyramidal neurons (immature and/or starting to mature) receive functional inputs from C-R cells axonic terminals within the first lamina (Chapter 5).

A montage of camera lucida drawings, from rapid Golgi preparations of 15-week-old fetuses, recapitulates, at scale, the human motor cortex basic cytoarchitectural organization, at this age (Fig. 3.8b). The first lamina C-R cells are characterized by numerous and closely arranged dendritic processes (a feature of embryonic C-R cells) and by long horizontal axonic processes with short ascending branches (Fig. 3.8b). The first lamina also contains the horizontal terminals of primordial corticipetal fibers and numerous radial glial terminal filaments with endfeet incorporated into the cortex EGLM, and the apical dendritic bouquets of all underlying gray matter pyramidal neurons as well as those from SP zone pyramidal-like neurons (Fig. 3.8b). Some of the deepest pyramidal neurons of the gray matter have developed short basal dendrites and a few proximal dendritic spines denoting their starting functional maturation (Fig. 3.8a). Their starting ascending functional maturation parallels the penetration of corticipetal fiber terminals of thalamic origin. Moreover, an intrinsic anastomotic capillary plexus has also started to form among the maturing neurons of P1 pyramidal cell stratum (Fig. 3.9). The primordial corticipetal fibers continue to cross the gray matter to reach the first lamina without establishing functional contacts with its pyramidal neurons (Fig. 3.8b). These primordial fibers target functionally the C-R cell horizontal dendrites and their presence precedes the formation and maturation of the gray matter PCP (Chapters 2 and 5).

At this gestational age, the human motor cortex is at a crucial transitional period characterized by two fundamental events. On the one hand, the PCP formation is nearly completed, some of its deepest and oldest P1 pyramidal neurons have begun their ascending functional maturation and the first gray matter intrinsic capillary anastomotic plexus has been established at this level (Figs. 3.6a–c, 3.7a–d, and 3.8a, b). On the other hand, some of the SP zone pyramidal-like and Martinotti neurons have begun to lose their original functional anchorages to the first lamina (Fig. 3.8b). At this age, the human motor cortex is starting to change its basic primordial functional organization based on the SP zone neurons in conjunction with the C-R cells into a new and definitive one based on its pyramidal neurons, also in conjunction with the first lamina C-R cells and specific thalamic corticipetal fibers. The descending axons of the deepest and oldest pyramidal neurons have already reached the WM and have become the principal source of corticofugal fibers leaving the maturing gray matter. From this gestational age forward, the basic functional organization of the human motor cortex will be, essentially, based on its pyramidal neurons, the thalamic fibers, and the C-R cells of first lamina. This functional transformation also represents a mammalian feature and innovation (Chapter 4).

At this age, the SP zone and its neurons have been displaced downward by the expansion of the pyramidal cell plate. Its pyramidal-like and Martinotti neurons, like C-R cells, have also undergo a significant developmental dilution, as their numbers, established at the start of development, will not increase further. The SP zone neurons, after losing their dendritic and/or axonic functional contacts with first lamina, undergo progressive retrograde changes and are progressively transformed into deep interstitial neurons. Deep interstitial neurons appear interspersed among the increasing number of fibers throughout the zone (Fig. 3.8b). The demarcation between the SP and the WM zones is no longer recognizable. The late functional role of these primordial SP zone neurons is unknown; they might represent link-up interneurons interrelating cortical, and subcortical centers.

At this age, the presence of horizontally migrating cells, throughout the motor cortex gray matter lower region, has been observed in some Golgi preparations (Fig. 3.8b). Although, their nature remains undetermined, they could represent horizontally migrating inhibitory interneurons that start to reach the developing motor cortex lower strata at this age. Their arrival and eventual functional incorporation into the motor cortex gray matter coincides with the functional maturation of its first pyramidal cell P1 stratum. The subsequent incorporation of additional inhibitory neurons will continue paralleling the ascending stratification and functional maturation of subsequent pyramidal cell strata (Chapter 6).

In view of the significance and possible relevance of the human motor cortex, at this prenatal age (and to further optimize the abundance of good Golgi preparations), a montage of camera lucida drawings has been made illustrating the entire thickness of the motor cortex as well as several features of its overall cytoarchitectural composition and organization (Fig. 3.9). The montage reproduces a perpendicular tissue block, from pial to ependymal surfaces, measuring 800 μm in width, 2,000 μm in height, and 150 μm in thickness. It recapitulates, at scale, the basic neuronal, fibrillary, microvascular and glial composition and organization of the human motor cortex, at 15th-week of gestation (Fig. 3.9). It roughly reproduces a region between two deep perforating vessels that cross the cortex from pial to paraventricular regions (Chapter 7). There are intrinsic anastomotic capillary plexuses at various levels of the motor cortex (Fig. 3.9, arrowheads). The most prominent ones are those through the paraventricular (PVZ), WM, SP zone and the newly established one through the gray matter first pyramidal cell P1 functional stratum (Fig. 3.9, arrowheads). Most perforating vessels descend through the gray matter without establishing interconnecting capillary plexuses until they reach its deep pyramidal cell P1 stratum. At this level, an anastomotic capillary plexus is formed between adjacent perforators throughout the developing motor cortex deep pyramidal cell stratum. Its formation coincides and parallels the starting functional maturation of its deepest and older pyramidal neurons. Some protoplasmic astrocytes have already reached this area and have established vascular endfeet with its intrinsic capillaries (Chapter 8). Above the pyramidal P1 stratum, the PCP pyramidal neurons remain immature and without an intrinsic capillary plexus. Numerous radial glial fibers cross the entire motor cortex from ependymal to pia surfaces. The demarcation between the SP and WM zones is no longer recognizable. At this age, the number of perforating vessel entering into the developing gray matter is greater (10–2) than those reaching the WM. The percentage of perforating vessels reaching the gray matter versus those reaching the WM continue to increase during late prenatal and postnatal cortical development, Eventually, the human cortex gray matter will become one of the most richly vascularized region of the body (Chapters 7 and 8).

The intrinsic anastomotic capillaries throughout the SP and the WM zones are already surrounded by numerous fibrous astrocytes with vascular endfeet. The number of fiber terminals throughout the SP and the WM zones has increase considerably. Also throughout the WM and matrix (M) zones there are groups of ascending neuronal precursors, with smooth leading processes, guided by radial glial filament. There are also numerous free-ascending glial precursors, without radial glial participation, with complex leading processes. The leading processes of some of these free-ascending glial precursors, as they reach the WM zone, bend in both directions accompany the increasing number of corticipetal and corticofugal fibers. Some of them could represent oligodendrocyte precursors (Marín-Padilla 1995). The paraventricular zone (PVZ) cytoarchitectural organization is quite complex with numerous free-ascending glial cell precursors with complex leading processes and by large tin-walled blood vessels covered by numerous glial processes (Fig. 3.9, thick arrows). This zone is also characterized by the presence of a thick bundle of horizontal fibers that run parallel to the ependymal surface for a considerable distance. The nature, origin, and functional target of this paraventricular fiber system are unknown to me. The ependymal surface (ES) has in addition to radial glial cell attachments other types of cells, still attached to its surface, with complex ascending processes, perhaps, representing glial cell progenitors still attached to the ependymal surface (Fig. 3.9).

3.4 Twenty-Week-Old Stage

Golgi preparations from seven fetuses, ranging from 20 to 26 weeks, represent this cortical development stage (Figs. 3.10 and 3.11).

At this age, the deep pyramidal neurons of stratum P1 have longer basal dendrites, some dendritic collaterals and more dendritic spines (Figs. 3.10a–d and 3.11). However, the distal regions of their apical dendrites remain smooth and spineless (Fig. 3.11). Their ascending maturation parallels the ascending penetration of corticipetal fiber terminals of thalamic origin and the establishment of an intrinsic capillary plexus in the region. The length of pyramidal neurons of stratum P1 is already 450 µm, nearly doubling the size of those of the motor cortex at 15th week of gestation. Some pyramidal neurons, above the pyramidal cell P1 stratum, have begun to develop short basal dendrites and a few proximal spines, denoting their starting ascending maturation as they have established functional contacts with ascending corticipetal fiber terminals of thalamic origin (Fig. 3.10a–d). Their starting maturation establishes the motor cortex second pyramidal cell functional P2 stratum (Figs. 3.10a–d and 3.11). Their apical dendrite distal region as well as all remaining PCP pyramidal neurons remained spineless and undifferentiated (Figs. 3.10a–d and 3.11). Although, silver deposits obscure the cytoarchitecture of the first lamina, thick horizontal axonic terminals of the C-R cells can be recognized in some Golgi preparations (Fig. 3.10c, d, arrows).

At this age, the functional maturation of P1 and P2 pyramidal cell strata has increased the thickness of the gray matter and further displaced downward the SP zone neurons. The identification of the displace SP zone primordial neurons becomes increasingly difficult among the increasing number of fiber terminals.

These neurons have been further diluted and their morphologic features have changed considerably. It is nearly impossible to distinguish the demarcation between the SP and the WM zones. Additional rapid Golgi preparations of the motor cortex deep regions will be necessary to view these transformed neurons. The photomicrographic magnification necessary to illustrate the gray matter neuronal cytoarchitectural organization will leave out of focus most of the SP zone (Fig. 3.10a–d).

A montage of camera lucida drawings, obtained from several Golgi preparations, recapitulates, to scale, the human motor cortex basic composition and cytoarchitectural organization at this developmental stage (Fig. 3.11). C-R cells (C-R) with typical early dendritic processes and their long horizontal axon terminals with short ascending branches are found scattered throughout the first lamina. Early primordial corticipetal fibers continue to ascend through the cortex without establishing functional contacts with the pyramidal neurons to terminate within the first lamina (I) into long horizontal terminals. At this age, the first lamina numerous horizontal fibers start to be segregate into upper and thinner ones and lower and thicker ones. The lower and thicker ones represent the C-R cells horizontal axonic terminals, while the thinner and upper ones represent the primordial corticipetal fibers horizontal terminals (Chapter 5). Both types of horizontal fibers can be followed for considerable distances within first lamina and their ends are not recognized. In Golgi preparations of older fetuses, the ascending functional maturation of the third pyramidal cell P3 stratum is starting. Some pyramidal neurons of this stratum have started to develop short basal dendrites (Fig. 3.10b, d). The remaining pyramidal neurons of the upper PCP are still immature with spineless apical dendrites, smooth bodies, and short descending axons. Vertical bundles of radial glial fibers cross the motor cortex and their terminal endfeet (g) are incorporated into the cortex expanding pial EGLM (Fig. 3.11). At this developmental stage, the motor cortex gray matter thickness, including differentiated and undifferentiated pyramidal neurons, ranges between 400 and 600 µm. It is important to point out that the increasing thickness of the gray matter is essentially due to the functional expansion of its lower pyramidal cell P1 and P2 functional strata. It should be pointed out that the thickness of the PCP upper undifferentiated zone remains essentially unchanged.

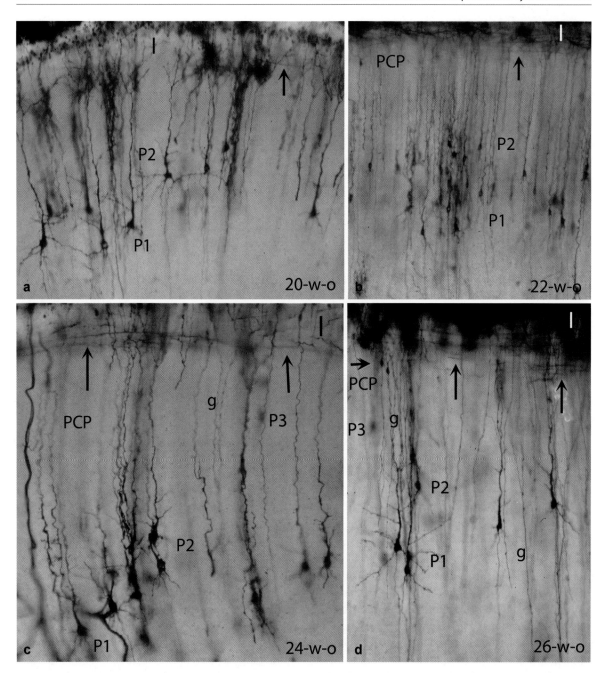

Fig. 3.10 Montage of rapid Golgi photomicrographs of the motor cortex of 20-, 22-, 24-, and 26- week-old human fetuses showing the degree of the PCP ascending stratification and the established P1 and P2 pyramidal cell strata (**a**, **c**) and the starting maturation of P3 pyramidal cell stratum. The PCP above the maturing P1–P3 pyramidal cell strata remains immature and undifferentiated (**d**). Despite silver deposits, in some (**a**, **c**, **d**) preparations, the C-R thick horizontal axonic terminals can be seen (*arrows*), crossing through the first lamina lower region

3.5 Thirty-Week-Old Stage

Four fetuses, ranging in age from 29 to 33 weeks, represent this prenatal developmental stage of the human motor cortex (Fig. 3.12). At this age, the human motor cortex is characterized by the ongoing functional maturation of at least four pyramidal cell P1–P4 strata (Fig. 3.12a–d).

Both, the gray matter cytoarchitectural complexity and the ascending functional maturation of its lower P1–P3 pyramidal cell functional strata have advanced considerably (Fig. 3.12a, b). The pyramidal neurons of

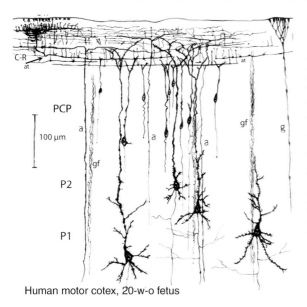

Fig. 3.11 Montage of camera lucida drawings, from rapid Golgi preparations, of the motor cortex of a 20-week-old human fetus summarizing, at scale, the basic cytoarchitectural organization of its main neuronal and fibrillar components. The maturing pyramidal neurons of P1 and P2 strata have longer basal dendrites, some collateral ones and increasing number of proximal apical dendritic spines, attesting their ongoing ascending functional maturation. The distal regions of the apical dendrites are still spineless. Dendritic spines are also developed in their terminal dendritic bouquets within the first lamina. The Cajal–Retzius neuron horizontal dendrites, axonic collaterals, descending axon, and thick horizontal axonic terminals coursing through the lower region are also illustrated. Ascending afferent fibers (a) crossing the PCP without making contacts with the pyramidal neurons and branching, into long horizontal terminals, within the first lamina, upper region, and some ascending radial glial (gf) fibers are also illustrated

the newly established pyramidal cell P4 functional stratum have developed short basal dendrites and a few proximal dendritic spines in the apical dendrites, which remain anchored to first lamina. The photomicrograph reproduced in Fig. 3.12b was taken from one of Cajal original rapid Golgi preparations, given to me by Prof. Fernando de Castro in 1967 (Chapter 12). The preparation label reads "infant motor cortex" without specifying the age. The preparation consists of several fragments of tissue, placed without any particular arrangement. The sections on this Golgi preparation have retained their original transparency, beauty, clarity, and authenticity, despite their old (110 years) age. Because at least four pyramidal cell (P1–P4) functional strata are recognized in this infant preparation, his age has been estimated to be around 33-weeks or perhaps older.

Although silver deposits obscure the first lamina, the presence of deep thick C-R cells horizontal axon terminals can be recognized in some preparations (Fig. 3.12b, d, arrows). At this age, some inhibitory interneurons start to be recognized among the lower pyramidal cell strata (Fig. 3.12d). Some Martinotti, Cajal double-bouquet and basket cells are recognized through the lower pyramidal cell strata (Fig. 3.12). Their morphological features and distinctive axonic terminals are not yet fully developed. The incorporation, recognition, and functional maturation of these inhibitory interneurons also follow an ascending progression paralleling the maturation of its associated pyramidal cell strata. The apical dendrites of pyramidal neurons of the PCP upper region, beneath the first lamina, remain immature and spineless.

3.6 Thirty-Five-Week-Old Stage

This developmental stage is represented by rapid Golgi preparations of three, 35-week-old fetuses. In one case, the quality of the rapid Golgi preparations was excellent, thus permitting a clear view of the ascending sequential maturation of various pyramidal cell strata at this age (Fig. 3.13).

At this age, the cytoarchitectural complexity of the human motor cortex gray matter is already extraordinary. The amount of information that a single Golgi preparation could carry is overwhelming (Fig. 3.13). In such a preparation, at least three sequential and superimposed levels (upper, middle, and lower) can be explored, each with different neurons, fibers terminals, blood capillaries, and glial cells. Months are needed to study a single one. A photomicrograph of a Golgi preparation (as the one illustrated in Fig. 3.13) only reproduces a single level of it. A complete description of the all information within a single Golgi preparation is beyond the scope of the present monograph. Consequently, only the ascending stratification and functional maturation of its gray matter pyramidal neurons will be described herein (Fig. 3.13), including some additional data concerning the recognition of some inhibitory interneurons and the extent of the ascending formation of the gray matter intrinsic microvascular and neuroglial systems.

At this age, the thickness of the motor cortex gray matter ranges between 700 and 900 μm, roughly representing the size (length) of the large pyramidal neurons of P1 stratum. The motor cortex shows the ascending functional maturation of five pyramidal cells strata: P1, P2, P3, P4, and P5 (Fig. 3.13). The

Fig. 3.12 Montage of photomicrographs from rapid Golgi preparations of the motor cortex of human fetuses aged 29- (**c**), 30- (**a, b**) and 32-week-old (**d**), respectively, showing its cytoarchitectural organization and composition and the degree of ascending stratification and maturation of the gray matter PCP. Four (P1, P2, P3, and P4) pyramidal cells functional strata are already recognized in the motor cortex at this age. The maturing pyramidal neurons are characterized by basal and collateral dendrites with spines and by an apical dendrite (often bifurcated) attached to first lamina by a terminal dendritic bouquet with spines only on its proximal segment. In some preparations the Cajal–Retzius cells thick horizontal axonic terminals are recognized running through first lamina lower zone (**a, d,** *arrows*). The photomicrograph reproduced in (**b**) was taken from one of Cajal old Golgi preparations, which despite its old age (prepared in 1899) remains clear, transparent, elegant, and authentic in demonstrating the morphology, size, distribution, and functional maturation of its pyramidal neurons (Chapter 12)

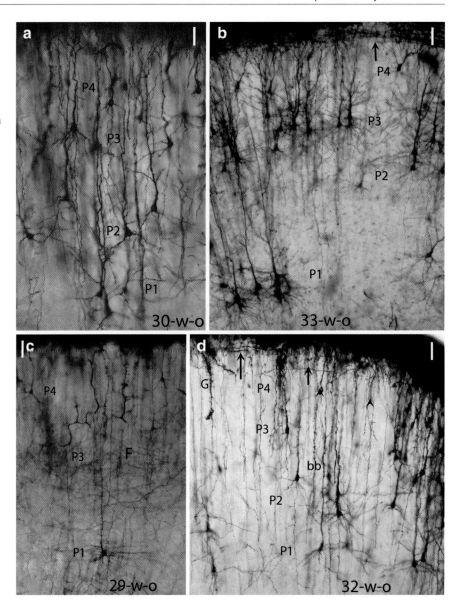

number of horizontal and vertical axonic terminals found among maturing pyramidal cells strata has increased considerably and some terminals make contacts with their dendritic spines. In addition, many additional inhibitory interneurons are already recognized, specially, through the lower pyramidal cell strata (Fig. 3.13). Through the four lower pyramidal cell strata (from P1 to P4), Martinotti, Cajal double-tufted and baskets cells are already recognized as well as other interneurons; morphologically less defined (Fig. 3.13). Below the pyramidal neurons P1 stratum, a few scattered polymorphous neurons are also recognized (Fig. 3.13). Short-linked anastomotic intrinsic capillary plexuses are recognized through all pyramidal cells strata paralleling their functional maturation (Chapter 7). A partial view of the gray matter intrinsic anastomotic capillary plexus can be seen in Fig. 3.13. Cataloguing all the information of a single rapid Golgi preparation of the motor cortex, at this age, will occupy an additional book.

The important point to emphasize is that, at this age, the starting functional maturation of an additional pyramidal cell P5 stratum is recognized in the human motor cortex (Fig. 3.13). Also, the structural and functional maturations of pyramidal neurons through P1, P2, P3, and P4 strata have increased considerably. This ongoing functional maturation is reflected in the increasing size of the pyramidal neurons, length of apical dendrites, length and number of basal and collaterals dendrites, and increasing number of dendritic spines (Chapter 4). At this age, an increasing number of axonic fiber terminals, from intra- and extracortical sources, fills the entire motor cortex and contributes to its advanced degree of functional maturation. Numerous axo-spino dendritic contacts are also recognized at this age (Chapter 4). At this age, it is simply impossible to adequately describe all the cytoarchitectural and functional features of the human motor cortex gray.

To appreciate the overall cytoarchitectural organization of the human motor cortex, the selected photomicrograph was taken at a low magnification (Fig. 3.13). While the illustration gains in perspective, it loses in details. It must be understood that with an increasing magnification the fine structural features of any cortical element will be better visualized and appreciated but the overall perspective will decrease (Chapter 4).

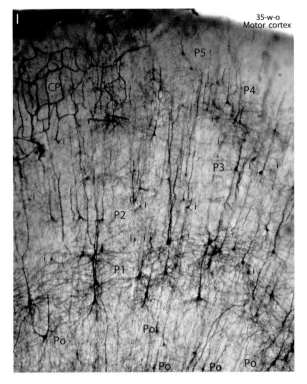

Fig. 3.13 Photomicrograph from a rapid Golgi preparation of the motor cortex of a 35-week-old human fetus showing a general view of its overall cytoarchitectural organization. In order to view the entire motor cortex, the photomicrograph was taken at a low microscopic magnification, which while gaining in perspective loses in details. At this age, the human motor cortex has already five, P1, P2, P3, P4, and P5, maturing pyramidal cell strata. Several deep polymorphous (Po) neurons as well as several non-pyramidal interneurons (i) are also illustrated. Due to the low photographic magnification their morphologic features are unclear. At this age, the motor cortex is crossed by innumerable horizontal and vertical fiber terminals, which are more prominent through the lower (P1,P2, and P3) pyramidal cell strata. A partial view of the anastomotic capillary plexus (CP) of the cortex intrinsic microvascular system is visible in the figure left upper cornet, which reproduces its tridimensional organization, the short interconnecting capillary loops and the small spaces between the capillaries where neuron reside. The amount of information that this preparation could offer is unimaginable. The Golgi preparation clarity and transparency permits, at a higher magnification, to explore and visualize in detail the morphology of its various elements, including the delicate morphology of the dendritic spines (Chapter 4)

3.7 Forty-Week-Old (Newborn) Stage

This developmental stage is represented by rapid Golgi preparations of the motor cortex of four newborn infants and by the good quality of some of them. To best illustrate the overall cytoarchitectural organization of the motor cortex of the newborn, the size and morphology of its various components and their complexity, I have chosen to use a montage of camera lucida drawings (made from rapid Golgi preparation) rather than a photomicrograph (Fig. 3.14a). A photo, at the low magnification required for viewing the entire gray matter, will result unclear and imprecise. On the other hand, a camera lucida drawing montage can illustrate, at scale, a simplified version of its basic cytoarchitectural organization and composition, which should be easier to visualize, interpret and understand. However, rapid Golgi photomicrographs of the newborn motor cortex can be seen in other figures that accompany the monograph (Figs. 4.1, 4.4, 4.9a, 6.1, and 9.3).

The newborn primary cortex motor, already about 2 mm thick, is characterized by the abundance of

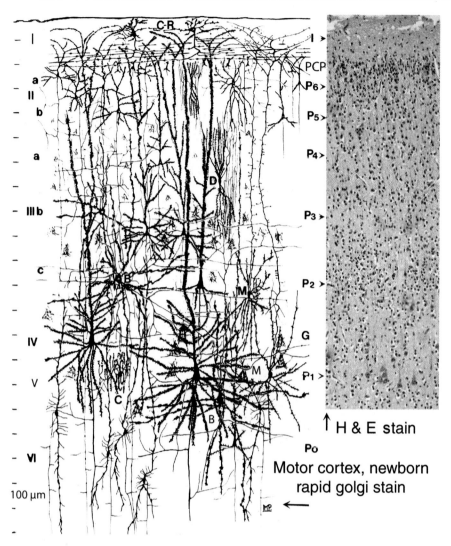

Fig. 3.14 Composite figure including a camera lucida drawing (**a**), from rapid Golgi preparations, and a H&E preparation (**b**) showing, at scale and comparatively, the overall cytoarchitectural organization and neuronal composition of the newborn infant motor cortex. Both the current (*left side*) and the proposed (*right side*) new developmental nomenclatures are also comparatively illustrated. Already, the newborn motor cortex has six – P1, P2, P3, P4, P5, and P6 – maturing pyramidal cell strata, which have been progressively and sequentially incorporated since the first P1 strata was established at 15 weeks of age. In addition, the newborn motor cortex (**a**, **b**) has a compact cellular zone, beneath first lamina, composed of undifferentiated small neurons that represents remnant of the original pyramidal cell plate (PCP). The neurons of this residual PCP will functionally mature during postnatal life that could incorporate an additional P7 pyramidal cell stratum to the human motor cortex. The decreasing sizes of pyramidal neurons from stratum P1 to –P6, the presence of basket (B), by-tufted (D), chandelier (C), polymorphous (Po) and Martinotti (M) neurons at various cortical levels are also illustrated. First lamina Cajal–Retzius (C-R) cells with their distinctive morphological features and long horizontal terminals are also illustrated. Compare the complex cytoarchitecture of the first lamina as revealed in Golgi preparations (A) with the apparent emptiness of this lamina in H&E (B) preparations. Other features are self- explanatory. The selection of a camera lucida drawing to illustrate the cytoarchitectural organization of motor cortex of the newborn is based on the fact that at this magnification, the revealing capabilities of a drawing are superior to those of a rapid Golgi preparation. However, rapid Golgi preparations of the motor cortex of newborn infants are included in the monograph, Figs. 4.1, 4.4, and 9.3

pyramidal neurons at all levels, with apical dendrites that reach and branch within the first lamina (see Fig. 9.3), innumerable vertical and horizontal fiber terminals through deep, middle, and upper strata (see Fig. 4.4), a variety of inhibitory interneurons (Chapter 6), a widespread intrinsic anastomotic capillary system (Chapter 7), and numerous protoplasmic astrocytes with vascular endfeet (Chapter 8). The pyramidal neurons of the newly established pyramidal cell P6 stratum are small, numerous, and already functionally mature with spines in both basal and apical dendrites (see Figs. 9.3 and 9.4). There are more pyramidal neurons in P6 stratum than in the rest of strata of the motor cortex. In addition, a thin band of residual PCP composed of small and still immature pyramidal neurons is also recognized, beneath the first lamina (Fig. 3.14a). This superficial residual PCP stratum is also recognized in H&E preparations of the newborn motor cortex (Fig. 3.14b). The functional maturation of this stratum will occur during early postnatal life. The eventual functional maturation of these PCP neurons could incorporate, postnatally, an additional pyramidal cell (P7?) stratum into the human motor cortex (Chapter 9). The motor cortex of all mammals, which I have been able to study also, has a residual PCP stratum of undifferentiated small pyramidal neurons, beneath the first lamina.

At birth, the human motor cortex gray matter thickness ranges from 1,500 to 1,700 μm (Fig. 3.14a, b). The size of its deeper and older pyramidal neurons of P1 stratum ranges between 1,500 and 1,600 μm, while that of superficial and younger ones of stratum P6 ranges around 100 μm (see Fig. 9.3). The newborn human motor cortex camera lucida montage (Fig. 3.14a) and the H&E preparation (Fig. 3.14b) reproduces comparatively, at a similar magnification, their respective cytoarchitectural organization and neuronal compositions. The monograph has introduced a new nomenclature that reflects the ascending and stratified (laminated) developmental maturation of the human motor cortex. The new nomenclature challenges, and should replace, the current one of descending (Layers I–VII) laminations, which remains unsupported by developmental data. To facilitate their correspondence, both nomenclatures, the descending laminar stratifications (left side) and the proposed one of ascending pyramidal cell strata (right side), are illustrated (Fig. 3.14a).

The camera lucida montage by reducing the number of elements illustrated, the motor cortex essential cytoarchitectural organization and composition can be best appreciated, visualized and understood (Fig. 3.14a). The camera drawings illustrate comparatively the stratification, size, location, distribution, and dendritic morphology of the pyramidal neurons, from stratum P1 up to P6. They also illustrate the size, dendritic, and axonic profiles of some of the accompanying inhibitory interneurons through various pyramidal cell strata, including: basket (B), double-tufted (D), Martinotti (M), and chandelier (C) cells, as well as deep polymorphous (Po) neurons (Fig. 3.14a). Other features illustrated in this camera lucida montage are self-explanatory. The extraordinary differences and revealing potentials between the rapid Golgi preparations obtained from camera lucida drawing montage (Fig. 3.14a) and the H&E preparations (Fig. 3.14b) are highlighted and emphasized. The inadequacy of routing H&E preparations in demonstrating the cerebral cortex various neuronal, fibrillary, synaptic, interneuronal, microvascular, and neuroglial systems is evident.

In conclusion, the human motor cortex prenatal development is characterized by the sequential ascending anatomical and functional stratifications of various pyramidal cell strata. It starts with the establishment of the first P1 pyramidal cell stratum (15-w-g), followed by the sequential ascending and stratified maturation of additional pyramidal cell strata P2 (20-w-g), P3 (26-w-g), P4 (30-w-g), P5 (35-w-g), and P6 (newborn infant). The additional pyramidal cell P6 functional stratum is unique to humans and will distinguish his motor cortex from that of other primates. Moreover, the newborn infant motor cortex (as well as that of other mammals) also has a thin band of residual undifferentiated PCP, which will mature functionally during postnatal life. This additional P7 pyramidal cell stratum could provide humans with unique motor capabilities and the control of new postnatally acquired motor abilities (motor creativity). It should also be emphasized that during the prenatal and postnatal developmental maturation the pyramidal neurons from all strata retain and increase their original functional contacts with first lamina C-R cell.

It is also important to point out that the incorporation of various types of inhibitory interneurons into the motor cortex also follows an ascending and stratified progression that parallels the ascending maturation of its various pyramidal cell strata (Chapter 4). Similarly, the prenatal development of the motor cortex intrinsic microvascular system follows an ascending and stratified sequence that also parallels the ascending maturation of its various pyramidal cell strata (Chapter 7). The incorporation of gray matter protoplasmic astrocytes into the motor cortex also parallels the ascending and stratified maturation of its various pyramidal cell strata as well as that of its intrinsic microvascular system, establishing numerous vascular endfeet with its capillaries (Chapter 8).

Some final commentaries may be appropriate The amount of cytoarchitectural information that a single good Golgi preparation, of the cerebral cortex, could offer is both extraordinary and fragmentary. That is the nature of the rapid Golgi procedure (Chapter 12). Any preparation may disclose some excitatory and inhibitory neurons as well as patches of its fiber, intrinsic microvasculature, and neuroglial system. Consequently, the study of a few Golgi preparations will be insufficient for acquiring any understanding of the cortex complex cytoarchitectural composition and organization. The only alternative is to make as many Golgi preparations as it may be possible and dedicate the necessary amount of time to study them. The number of rapid Golgi preparations, on which the present monograph is based, is around 4,500 single cuts and the amount of time dedicated to their study has been, so far, 36 years. Although not all preparations are of a good revealing quality, the large number of them studied compensates for their deficiencies. In my experience, any Golgi preparation has always some information to offer, even if it is fragmentary and incomplete. During the years dedicated to study these preparations, I have observed much more that I have been able of interpreting, understanding and/or explaining. Consequently, my attention has been concentrated in exploring the developmental histories of some of the human motor cortex basic components, including: its essential pyramidal neurons (Chapter 3 and 4), the first lamina (Chapter 5), some excitatory-inhibitory neuronal systems (Chapter 6), the intrinsic microvascular system (Chapter 7) and the gray matter special protoplasmic astrocytes (Chapter 8). The monograph introduces a new unifying developmental cytoarchitectonics theory, applicable to all mammalian species, and of a new nomenclature (Chapter 9).

In addition to its large size and larger number of neurons, what actually characterizes and distinguishes the human brain from that of other mammals is the number of functional pyramidal cells strata established in it. The astonishing motor capabilities of the human hand (liberated by Bipedism) are assembled for action within the motor cortex. From an evolutionary perspective, the human hand extraordinary motor potentials may have been the force behind the formation of an additional P6 pyramidal cell stratum in his motor cortex. No other primate, including the chimpanzee, has such capable hands and consequently the cytoarchitectural organization of their motor cortex should be simpler and the number of pyramidal cell strata should reflect their motor capabilities. Primates seem to accomplish all their motor needs and activities with five pyramidal cell strata as the present monograph proposes (Chapter 9). During mammalian evolution, as the animal motor needs and capabilities increase, the number of functional pyramidal cell strata formed in their motor cortex increases accordingly.

The large P1 pyramidal neurons represent the motor cortex main functional outlet. Their final functional action seems to be the result of descending functional inputs from all pyramidal neurons of the above strata (Chapter 4). The complexity of their actions should reflect the number of supporting and participating pyramidal cell strata. The present developmental study proposes that the number of pyramidal cell strata established in the motor cortex reflects the motor capabilities of each mammalian species. This new conception challenges the old Darwinian idea, supported by classic neuroscientists and universally accepted today, which proposes that the only difference in the cerebral cortex among different mammalian species is of degree but not of kind.

The present chapter describes the developmental history of the human motor cortex essential and fundamental pyramidal neurons. It proposes that the function of the remaining cortical elements contributes to that of the pyramidal neurons and that their developmental histories are concomitant and parallel. There are additional less-defined neuronal systems and types of neurons throughout the cerebral motor cortex that remain unexplored, although they may also contribute to the overall function of the pyramidal neurons. Moreover, there are other regions throughout the human cerebral cortex that remain essentially unexplored, and their developmental histories unknown. Consequently, the observations presented in this monograph are incomplete and should be considered as stepping stones for additional studies (hopefully using the rapid Golgi procedure), which will be needed before we can achieve a more complete understanding of the cytoarchitectural and functional organizations of the human cerebral cortex and, in general, of that of any other mammalian species.

References

Bayer SA, Altman J (1991) Neocortical development. Raven, New York
Blinkov SM, Glezer II (1968) The human brain in figures and tables. Plenum Press, New York
Cajal SR (1899a) Estudios sobre la corteza cerebral humana. Estructura de la cortíza motríz del hombre y mamíferos. Revista Trimestral Micrográfica 4:117–200
Cajal SR (1899b) Estudios sobre la corteza cerebral humana. Corteza visual. Revista Trimestral Micrográfica 4:1–63
Cajal SR (1900a) Estudios sobre la corteza cerebral humana. Estructura de la corteza acústica. Revista Trimestral Micrográfica 5:129–183
Cajal SR (1900b) Estudios sobre la corteza cerebral humana. Estructura de la corteza cerebral olfativa del hombre y mamíferos. Revista Trimestral Micrográfica 6:1–150
Kölliker AV (1896) Handbuch der Gewebelebre des Menschen. Bd. II, Nervensystem des Menschen und der Tiere. Leipzig, Engelmann
Larroche J-C (1977) Developmental pathology of the neonate. Excepta Medica, Amsterdam
Larroche J-C (1981) The marginal layer in the neocortex of a 7-month-old human embryo. Anat Embryol 162:301–312
Larsen WJ (1993) Human embryology. Churchill Livingstone, New York
Luskin ML, Shatz CJ (1985) Studies of the early generated cells of the cat's visual cortex: cogeneration of subplate and marginal zones. J Neurosci 5:1062–1075
Marín-Padilla M (1967) Number and distribution of the apical dendritic spines of the layer V pyramidal neurons in man. J Comp Neurol 131:475–490 (This study was carried out using Cajal old Golgi preparations)
Marín-Padilla M (1968) Cortical axo-spinodendritic synapses in man. Brain Res 8:196–200
Marín-Padilla M (1969a) Origin of the pericellular baskets of the pyramidal cells of the human motor cortex. Brain Res 14:633–646
Marín-Padilla M, Stibitz GR, Almy CP, Brown HN (1969b) Spine distribution of the layer V pyramidal cells in man. Brain Res 12:493–496
Marín-Padilla M (1970a) Prenatal and early postnatal ontogenesis of the human motor cortex. A Golgi study. I. The sequential development of the cortical layers. Brain Res 23:167–83
Marín-Padilla M (1970b) Prenatal and early postnatal ontogenesis of the human motor cortex. A Golgi study. II. The basket-pyramidal system. Brain Res 23:185–191
Marín-Padilla M (1971) Early prenatal ontogenesis of the cerebral cortex (neocortex) of the cat (Felis domestica). A Golgi study. I. The primordial neocortical organization. Zeitschrift für Anatomie und Entwicklungeschichte 134:117–145
Marín-Padilla M (1972a) Prenatal ontogenetic history of the principal neurons of the neocortex of the cat (Felis domestica). A Golgi study. II. Developmental differences and their significance. Zeitschrift für Anatomie und Entwicklungeschichte 136:125–142
Marín-Padilla M (1972c) Double origin of the pericellular baskets of pyramidal cells of the human motor cortex. Brain Res 38:1–12
Marín-Padilla M (1972b) Structural abnormalities of the cerebral cortex in human chromosomal aberrations. A Golgi study. Brain Res 44:625–629
Marín-Padilla M (1974a) Structural organization of the cerebral cortex (motor area) in human chromosomal aberrations. A Golgi study. I. D1 (13-15) trisomy. Patau syndrome. Brain Res 66:375–391
Marín-Padilla M (1974b) Three-dimensional reconstruction of the basket of the human motor cortex. Brain Res 70:511–514
Marín-Padilla M (1976) Pyramidal cell abnormalities in the motor cortex of a child with Down's syndrome. A Golgi study. J Comp Neurol 167:63–82
Marín-Padilla M (1978) Dual origin of the mammalian neocortex and evolution of the cortical plate. Anat Embryol 152:109–126
Marín-Padilla M (1983) Structural organization of the human cerebral cortex prior to the appearance of the cortical plate. Anat Embryol 186:21–40
Marín-Padilla M (1985a) Neurogenesis of the climbing fibers in the human cerebellum; A Golgi study. J Comp Neurol 235:82–96
Marín-Padilla M (1985b) Early vascularization of the embryonic cerebral cortex: Golgi and electron microscopic study. J Comp Neurol 241:237–249
Marín-Padilla M (1987a) The Golgi Method. In: Adelman G (ed) Encyclopedia of neurosciences, vol 1. Birkhauser Boston, Cambridge, MA, pp 470–471
Marín-Padilla M (1987b) The chandelier cell of the human visual cortex. A Golgi study. J Comp Neurol 265:61–70
Marín-Padilla M (1990a) Three-dimensional structural organization of layer I of the human cerebral cortex. A Golgi study. J Comp Neurol 229:89–105
Marín-Padilla M (1990b) The pyramidal cell and its local-circuit interneurons. A hypothetical unit of the mammalian cerebral cortex. J Cogn Neurosci 2:180–194
Marín-Padilla M (1992) Ontogenesis of the pyramidal cell of the mammalian neocortex and developmental cytoarchitectonics: a unifying theory. J Comp Neurol 321:223–240
Marín-Padilla M (1995) Prenatal development of fibrous (white matter), protoplasmic (gray matter), and layer I astrocytes in the human cerebral cortex. A Golgi study. J Comp Neurol 358:1–19
Marín-Padilla M (1996) Developmental neuropathology and impact of perinatal brain damage. I. Hemorrhagic lesion of neocortex. J Neuropathol Exp Neurol 55:746–762
Marín-Padilla M (1997) Developmental neuropathology and impact of perinatal brain damage. II. White matter lesions of the neocortex. J Neuropathol Exp Neurol 56:219–235
Marín-Padilla M (1998) Cajal-Retzius cell and the development o the neocortex. Trends Neurosci (TINS) 21:64–71
Marín-Padilla M (1999) Developmental neuropathology and impact of perinatal brain damage. III. Gray matter lesions of the neocortex. J Neuropathol Exp Neurol 58:407–429
Marín-Padilla M (2000) Perinatal brain damage, cortical reorganization (acquired cortical dysplasia), and epilepsy.

In: Williamson P, Siegel AM, Roberts D, Vijay VM, Gazzaniga M (eds) Neocortical epilepsies. Lippincott Williams & Wilkins, Philadelphia, PA, pp 156–172

Marín-Padilla M, Marín-Padilla T (1982) Origin, prenatal development and structural organization of layer I of the human cerebral (motor) cortex. Anat Embryol 164:161–206

Marín-Padilla M, Parisi JE, Armstrong DL, Kaplan JA (2002) Shaken infant syndrome. Developmental neuropathology, progressive cortical dysplasia and epilepsy. Acta Neuropathologica 103:321–332

Mrzljak LH, Uylings BM, Kostovic I, van Eden CG (1988) Prenatal development of neurons in the human prefrontal cortex. I. A quantitative study. J Comp Neurol 277: 355–386

O'Rahilly R, Müller F (1994) The embryonic human brain: an atlas of developmental stages. Wiley-Liss, New York

Poljakow GI (1961) Some results of research into the development of the neural structure of the cortical end of the analyzer in man. J Comp Neurol 117:197–212

Retzius G (1893) Die Cajalschen Zellen der Grosshirnrinde beim Menschen und bei Såugetieren. Biologie Unters 5:1–9

Retzius G (1894) Weitere Beitráge zur Kenntnis der Cajalschen Zellen der Grosshrnrinde des Menschen. Biologie Unters 93:29–34

The Mammalian Pyramidal Neuron: Development, Structure, and Function

In all mammalian species, the cerebral cortex (neocortex) is characterized and distinguished by the abundance of pyramidal cells, representing roughly 75% of its neurons. This neuronal type represents a mammalian innovation and is characterized by unique developmental, morphological, and functional features (Marín-Padilla 1971, 1992). It also represents the essential functional outlet of the neocortex, such that the remaining cortical neurons contribute, directly and/or indirectly, to its functional role. The prenatal development and functional maturation of pyramidal neurons is a sequential, ascending, and stratified process (Marín-Padilla 1992, 1998). To comprehend why the mammalian neocortex is stratified (laminated), the developmental, structural, and functional features of the pyramidal neuron must be first clearly understood. The present chapter proposes that the number of pyramidal cell's functional strata (laminations) established in the cerebral cortex varies among different mammalian species, increases in the course of phylogeny, and reflects the animal motor abilities and capabilities. The essential developmental, morphological, and functional features of this uniquely mammalian neuron are explored in the present chapter.

In this chapter, the developmental term cortical plate (CP), has been substituted herein by the term pyramidal cell plate (PCP) for the following reason. From the start of development to the adult brain, this cortical formation is essentially composed of pyramidal neurons of different sizes, all functionally anchored to first lamina by their apical dendrite (Chapter 3). In the course of cortical developments, local circuit inhibitory and other non-pyramidal neurons become incorporated into the PCP paralleling the ascending maturation of its various pyramidal cell strata. In the newborn motor cortex, Martinotti, double-tufted, baskets and chandelier inhibitory interneurons are recognized in all pyramidal cell functional strata (Fairen et al. 1984; Jones and Hendry 1984; Marín-Padilla 1969, 1970b, 1974b, c, 1987, 1990b; Marín-Padilla and Stibitz 1974; Peters 1984; Peters and Saint Marie 1984; Somogyi and Cowey 1984). The developmental history of the inhibitory interneurons of the human motor cortex is described in the fifth Chapter 5. Non-pyramidal cortical neurons, lacking dendritic functional attachment to the first lamina of variable sizes and morphologies, have also been described in the cerebral cortex (Fairen et al. 1984). The developmental histories and incorporation time of non-pyramidal neurons remain unexplored.

To describe the developmental, morphological, and functional features of the mammalian pyramidal neuron, the human motor cortex neurons are used as models. Of these, the pyramidal neurons of stratum P1 (layer V in current nomenclature) are selected (Figs. 4.1, 4.2a–c and 4.3). The essential features of these neurons (described below) should be applicable to the pyramidal neurons of other cortical strata as well as to those of other cortical regions. And, in general terms, they should also be applicable to pyramidal neurons of the motor cortex in other mammalian species.

4.1 Developmental Features

All migrating neuroblasts, which build up the neocortex PCP, attracted by reelin from Cajal–Retzius (C-R) cells and guided by radial glia fibers, must reach the first lamina. Upon reaching it, they lose the glia attachment and become pyramidal neurons with an

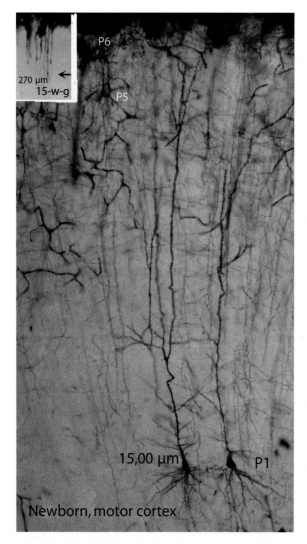

Fig. 4.1 Montage of photomicrographs from rapid Golgi preparations showing, comparatively, similar magnification, the size and functional maturation of a pyramidal neuron of P1 stratum (layer V of current nomenclature) from the motor cortex of a 15-week-old fetus (*inset*) and of a newborn infant. The extraordinary prenatal developmental morphological and functional maturation of this essential neuron of the human motor cortex has not been fully appreciated and/or recognized. The pyramidal cell size of the 15th-week-old fetus is 270 μm, has very short basal dendrites, and a couple of dendritic spines, while that of the newborn infant is 1,500 μm and all its dendrites are covered by spines. Only the rapid Golgi procedure is capable of demonstrating these neurons extraordinary, progressive, ascending, and stratified developmental morphological, and functional maturations (see Chapter 3)

apical dendrite that branches within the first lamina and establish functional contacts with C-R cells axon terminals (Marín-Padilla and Marín-Padilla 1982; Marín-Padilla 1992, 1998). All pyramidal neurons of the neocortex are generated in the ependymal epithelium. In humans, the PCP formation starts around the 7th week of gestation and is nearly completed by the 15th week of gestation (w-g). The PCP neurons start their ascending and stratified functional maturation around the 15th week of gestation and will continue through both prenatal and early postnatal life (Chapter 3). At 15-week gestation, the motor cortex PCP is a 100 cell thick plate extending throughout the entire neocortex, composed of tightly packed immature pyramidal neurons of various sizes, all functionally anchored to first lamina by their apical dendrite. At this age, no other neuronal types are recognized in the PCP. From this gestational age up to the time of birth and postnatally, the motor cortex pyramidal neurons mature anatomically and functionally, following an ascending sequential stratification, without losing their first lamina attachment and/or their body placement or cortical depth (Chapter 3).

The developmental requirement for all migrating neuroblasts of reaching the first lamina explains their inside–outside placement within the PCP. Any migrating neuroblast must by-pass all preceding ones in order to reach the first lamina such that all preceding pyramidal neurons – to retain their original functional anchorage to first lamina – must elongate "upwardly" their apical dendrite. Pyramidal neurons evolving from early arriving neuroblasts will occupy the PCP lower strata and will have the longest apical dendrites; while those arriving last will occupy the upper strata and will have shortest apical dendrites. The length of apical dendrites of pyramidal neurons located between the oldest and youngest ones will be intermediate. Consequently, the length of the apical dendrites denotes the arrival time of the pyramidal neuron at first lamina and its developmental age. Regardless of their morphological appearance, neurons that lose their functional attachment to first lamina (subplate neurons) and/or fail to develop them should not be considered as genuine pyramidal neurons (Marín-Padilla 1992).

The mammalian neocortex is composed of superimposed and functionally interconnected pyramidal cell strata ranging from lower and older ones to superficial and younger ones. Therefore, the mammalian neocortex cytoarchitectural organization should be understood not as a descending (up–down) series of laminations (layers I, II, III, IV, V, and VI) as is portrayed in the current nomenclature, but by a series of

4.1 Developmental Features

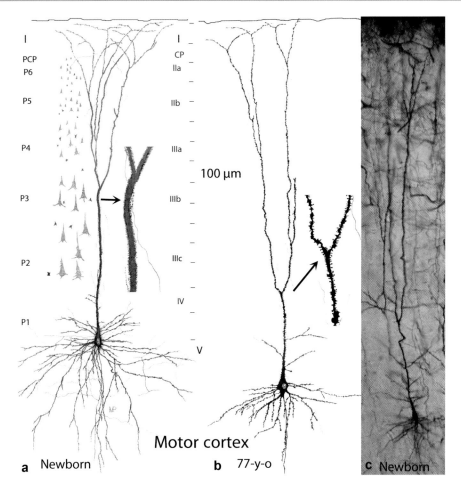

Fig. 4.2 Montage of camera lucida drawings and a photomicrograph (**c**) from Golgi preparations showing, comparatively, the size, morphology, functional maturation, dendritic profiles, and the number of dendritic spines of P1 pyramidal neurons from the motor cortex of a newborn infant (**a**) and a 77-year-old man (**b**). The insets (**a**) reproduce closer views of the neurons apical dendrites showing their variable morphology, number of dendritic spines, and some axo-spino-dendritic contacts with axon terminals. The old man's pyramidal neuron (**b**) has thinner dendrites, fewer dendritic branches, and a reduced number of dendritic spines, also some spines are structurally abnormal. Both the current (*right*) and the proposed (*left*) developmental nomenclature are shown comparatively (**a**). The rapid Golgi photomicrograph (**c**) represents a perpendicular section of a newborn motor cortex showing, a nearly complete P1 pyramidal neuron, at roughly the same magnification of those of the drawings for comparison. Scale: 100 μm

ascending stratified interconnected pyramidal cells (P1, P2, P3, P4, P5, and P6) functional strata. Both the current and the proposed developmental nomenclatures are shown comparatively in some of the illustrations (Fig. 4.2a and Fig. 3.14).

From the start of development, any pyramidal neuron is locked into position between its functional anchorage to first lamina and the cortical depth (original placement) of its soma (Fig. 4.3). Consequently, during their prenatal development, all pyramidal neurons must elongate "upwardly" their apical dendrite without losing either anchorage (Fig. 4.3). During early development stages, all pyramidal neurons elongate their apical dendrite anatomically to accommodate the subsequent incorporation, into the PCP, of additional ones. Later in development, all pyramidal neurons elongate their apical dendrites functionally (physiologically) by adding postsynaptic membrane to their main shaft, for the formation of dendritic spines and other types of synaptic contacts (Fig. 4.3, inset). This progressive functional elongation is also applicable to the neuron basal, collateral, and terminal dendrites within the first lamina (Fig. 4.3).

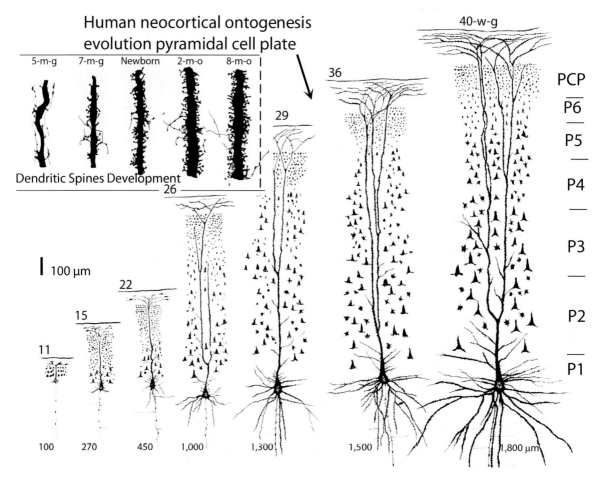

Fig. 4.3 Montages of camera lucida drawings, from Golgi preparations, showing, in the inset, the morphology and the increasing number of spines (*inset*) in the apical dendrites of P1 pyramidal neurons, of fetuses, and infant's motor cortex, from the 5th week of gestation to the 8th postnatal month; and, in the large drawing, a schematic representation of the progressive ascending anatomical and functional maturations of P1 pyramidal neurons of the human motor cortex, from the 11th to the 40th week of gestation. Also illustrated are the ascending functional maturation and stratification of the motor cortex PCP and the proposed new developmental nomenclature. Other aspects of the figure are self-explanatory. Scale: 100 μm

Both types of developmental growths (elongations) invalidate a generally held idea, which proposes that apical dendrites of the pyramidal neurons grow up from cell body to the first lamina. Considering the extraordinary and increasing structural complexity of the neocortex, it is difficult to visualize the long apical dendrites of these neurons could grow (elongate) perpendicular and so clean up to the pial surface, On the other hand, since all pyramidal neurons retain their original first lamina dendritic attachment, as the cortex thickness increases, all apical dendrites must elongate upward (Fig. 4.3). This new conception is applicable to all pyramidal neurons throughout the mammalian neocortex regardless of the size or location, including those within deep sulci and/or superficial gyri. It also explains why even dendritic bifurcations (a common feature among large pyramidal neurons) are also perpendicular to the pial surface (Figs. 4.1–4.3).

4.2 Morphological Features

To best describe the unique morphologic features of the mammalian pyramidal neuron, the newborn motor cortex giant pyramidal neuron (Betz cell) of stratum P1 (layer V in current nomenclature) is used as a model (Figs. 4.1–4.3). It is the larger and older pyramidal neuron of the motor cortex as well as its main functional outlet. In the human motor cortex, it is characterized by a large and roughly triangular body with long basal dendrites and by an ascending apical dendrite, with or without bifurcation, with collateral branches that cross the entire thickness of the cortex and terminates forming a dendritic bouquet within the first lamina (Figs. 4.1–4.3). At the 15th week of gestation, the deep and older pyramidal neurons in the motor cortex have an average length of about 270 μm, a few short basal dendrites, and a couple of proximal apical dendritic spines (Figs. 4.1 inset, 4.8b). The starting functional maturation of these neurons, documented by the development of short basal dendrites and a few dendritic spines, denotes the beginning of the motor cortex PCP ascending functional maturation and the establishment of its first pyramidal cell P1 (layer V in current nomenclature) functional stratum. In the newborn motor cortex, the average height of a P1 pyramidal neuron is about 1,500 μm, has long basal and collateral dendrites, an apical dendrite (with or without bifurcations) that crosses the cortex entire thickness and branches into a terminal dendritic bouquet within first lamina (Figs. 4.1 and 4.2a, c). And thousands of dendritic spines already cover its apical, basal, collateral, and terminal dendrites (Marín-Padilla 1967). In addition, this pyramidal neuron receives countless direct axo-dendritic synaptic contacts along its dendrites as well as specific synaptic terminals from its associated inhibitory interneurons (Chapter 6). The extraordinary anatomical and functional growths of the human motor cortex PI pyramidal neurons, between the 15th week of gestation and birth is comparatively illustrated in Fig. 4.1. The prenatal developmental history of the other pyramidal neurons strata of the human motor cortex has been explored in this Chapter 3.

In the developing cat motor cortex, the apical dendrite of a P1 pyramidal neuron elongates anatomically from 25 μm, at 27 days to 600 μm by the time of birth. And physiologically (functional maturation) from a few apical dendritic spines, at 35-days to nearly 700 μm, by birth time (Marín-Padilla 1972, 1992). Similar anatomical and functional growths are applicable to P1 pyramidal neurons of the motor cortex of mice and hamsters. This dual anatomical and functional growth is equally applicable to pyramidal neurons of all strata of the mammalian motor cortex.

In general terms, aging seems to affect pyramidal neurons both anatomically and functionally. The number, thickness, and length of their dendrites decrease as well as the number of dendritic spines and of other types of synaptic contacts (Fig. 4.2b). Probably, the synaptic contacts from intrinsic and extrinsic sources also decrease with aging. Further studies are needed to clarify these age-related anatomical and functional deficiencies as they may be related to senility and dementia.

4.3 Functional Features

Throughout the motor cortex, the dendrites of all pyramidal neurons are, in fact, immerged into a sea of horizontal, vertical and oblique axonic terminals (Fig. 4.4). The large number of fiber terminals illustrated in this figure only represents those visible within a single microscopic view of a much thicker – between 150 and 250 μm – rapid Golgi preparation. The number of fiber terminals through an entire Golgi preparation is unimaginable. Rapid Golgi preparations occasionally provide unexpected views of the cortex cytoarchitectural organization of exquisite beauty, clarity, and authenticity (Fig. 4.4). Compare this figure with Fig. 9.3. Both photomicrographs are from rapid Golgi preparations of newborn infants' motor cortex. While, the first figure illustrates the innumerable fiber terminals that surround the apical dendrites of the pyramidal neurons with only a few of them stained, the second one illustrates several apical dendrites of the pyramidal neurons crossing the cortex with only a few fiber terminals stained. It should be pointed out, that if all fiber terminals and apical dendrites were stained on either preparation, they would be impossible to study, unravel, and/or comprehend. It must also be understood that using the rapid Golgi procedure these results are impossible to predict and/or to plan ahead as well as obtain any other type of desired result. The procedure mystique is that it can offer unexpected, unique, clear, extraordinary, and authentic views of the nervous tissue organization. What any Golgi preparation offers, even incomplete, must be accepted, carefully studied, and recorded. The aim should be to prepare and study, as many preparations as it may be possible, hoping that some of them will offer those exceptional and incomparable views, which might contribute to solving part of the puzzle that represents the mammalian neocortex cytoarchitectural organization.

During their functional maturation, pyramidal neurons receive ascending and sequential synaptic contacts

Fig. 4.4 A rapid Golgi photomicrograph showing the extraordinary and complex cytoarchitectural and fibrillar (F) organizations from a newborn infant motor cortex. The dendrites of all pyramidal and non-pyramidal cortical neurons are immerged into a sea of innumerable vertical, transverse, and oblique fiber terminals, which extends throughout the cortex entire thickness. The illustration reflects the already remarkable and startling cytoarchitectural complexity of a human newborn motor cortex. Moreover, the large number of fiber terminals shown in the photo are simply those viewed in a single microscopic sight of a much thicker Golgi preparation. Consequently, the actual number of fibers terminals and synaptic contacts throughout the newborn motor cortex is unimaginable and both difficult to categorize and to comprehend. Moreover, most cortical synapses are not stable contacts but constantly modifying ones. The photo also illustrates some P1 pyramidal neurons (P1), several apical dendrites, and the presence of other neuronal types, including: deep polymorphous (Po), basket (B), and double- tufted (DT) neurons (compare this figure with Fig. 9.3)

from an increasing number of specific and nonspecific thalamic afferent (corticipetal) fibers as well as from interhemispheric (callosal) and cortico-cortical fiber terminals. In addition, pyramidal neurons receive contacts from intrinsic neuronal sources, including those from excitatory and inhibitory neuron. The dendritic spines are their main receptors for excitatory synaptic inputs (Figs. 4.5–4.7). Cajal first described these fine dendritic excrescences in cortical pyramidal neurons and later in cerebellar Purkinje cells (Cajal 1891, 1894, 1896). For a long time, the anatomical and functional nature of the spine was the source of controversy as many consider them as mere silver artifacts (Cajal 1923). Cajal insisted in their anatomical reality and functional relevance pointing out that spines will significantly increase the dendrite receptive surface and will facilitate direct contacts with adjacent fiber terminals (Cajal 1891, 1894, 1896, 1911). In 1959, Gray, using the electron microscope, demonstrated the spine anatomy and the functional contact on its terminal round head (postsynaptic structure) with an axon terminal (Gray 1959). In rapid Golgi preparation, the dendritic spines, at a low microscopic magnification, appear indeed as fussy dendritic excrescences, which could easily be mistaken as artifacts (Fig. 4.5a). However, at medium and/or higher magnifications, the spine distinct anatomical features are clearly recognized (Fig. 4.5b, c). Spines emerge from all sides of the dendrite, have a short fine pedicle, and terminate forming a small enlargement, which represents its postsynaptic apparatus.

In Golgi preparations, spines are seen in profile and, therefore, their length and morphology are quite variable (Figs. 4.3, insets and 4.5c). Only spines arising from the dendrite sides can be seen entirely. Those arising from the dendrite anterior and/or posterior walls are only partially seen and/or not seen at all. And, they may appear as short and/or deformed spines (Figs 4.5a–c and 4.6a–d). The spine many overlap resulting in thicker and/or abnormal appearing one. While observing dendritic spines in rapid Golgi preparations, it is difficult to evade the thoughts that these structures are constantly moving and modifying their size, length, and terminal head (Figs. 4.5b, c and 4.6b–e). During development, dendritic spines are being constantly formed, reabsorbed, and reformed, in response to variations of their functional contacts. Further reviewing the dendritic spine features is beyond the scope of the present chapter. Several recent studies have explored, in depth, the dendritic spine morphology, synaptic neurotransmitters, functional features, as well as their structural and functional modifications (Shepherd 1996; De Felipe 2006; Arrellano et al. 2007).

Fig. 4.5 Montage of photomicrographs illustrating the morphological features of the dendritic spines through middle segments of the apical dendrites of large P1 pyramidal neurons, from rapid Golgi preparations, of newborns motor cortex. The dendritic spines are seen at low (**a**), medium (**b**), and high (**c**) magnifications. Axo-spino dendritic synaptic contacts between axon terminals and spines (**b**, **c**, *arrows*) are also visible. The blood capillary (bc) diameter (**c**) ranges between 5 and 6 μm, which can be used comparatively to determine those of dendrites and dendritic spines. The inset (C) illustrates at a higher magnification the variable morphology of dendritic spines as well as the profile of two lateral spines (*arrow heads*) with round terminal heads. The spines' variable shapes and lengths result from viewing them from different perspectives and by overlapping. Fiber terminals (F) are also illustrated. Scale: 5 μm

Dendritic spines represent the main receptive system of the neuron for excitatory synapses. The number of dendritic spines, along the apical dendrite of pyramidal neurons, increases progressively from the 15th week of gestation to birth time (Fig. 4.3, inset). By birth time, thousand of dendritic spines cover the pyramidal neurons dendrites (Marín-Padilla 1967). The spine distribution along the apical dendrite follows a distinct pattern, which is similar in all mammalian species investigated (Marin-Padilla 1967; Valverde 1967; Marín-Padilla and Stibitz 1968; Marín-Padilla et al. 1969). During postnatal development, the number of spines continues to increase and to undergo significant modifications in response to intracortical intrinsic and extrinsic input variations as well as from extracortical sources.

A pyramidal neuron also receives direct synaptic contacts on their dendritic shaft (axo-dendritic synapses) from both intrinsic and extrinsic neuronal sources. The number and types of neuronal sources responsible for the direct axo-dendritic synapses contact on pyramidal neurons remain poorly studied and understood. A pyramidal neuron also receives specific synaptic terminals from its associate inhibitory interneurons (Marín-Padilla 1969, 1974, 1987, 1990). The Martinotti cell axon seems to target the terminal

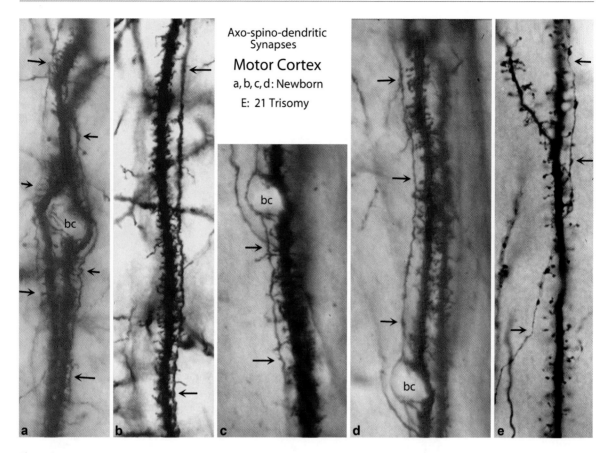

Fig. 4.6 Montage of rapid Golgi photomicrographs showing various (a–e) examples of multiple parallel axo-spino-dendritic synaptic contacts (arrows). Despite their possible functional relevance, this type of multiple synaptic spine contacts by the same axon terminal has been seldom described in the literature. In these multiple synaptic contacts the relationship between axon terminal and the dendritic spines varies: in some (b) the axon terminal simply parallel de dendrite, in others (a) both axon terminal and dendrite bend around a blood capillary, in still others (c, d) the axon bend on one side of a blood capillary and the dendrite on the other, while in others (e) the axon goes behind the dendrite making contact with the spines on both sides of the dendrite. This last case is from the motor cortex of an infant with 21-trisomy. The number of synaptic contacts established between an axon terminal and the spines of a dendrite could be significant (c, d), perhaps attesting to their functional importance

dendritic bouquet of pyramidal neurons of its cortical stratum. These inhibitory neuron's ascending axon reaches the first lamina and branches into a bouquet that mimics the terminal dendrites of the pyramidal neurons. The basket cells target the pyramidal neuron soma forming a complex pericellular basket with numerous axo-somatic synaptic contacts. Double-tufted interneuron seems to target the neuron apical dendritic shaft forming direct axo-dendritic synaptic contacts. These interneurons axon branch into several ascending and descending terminals, which parallel the apical dendrites of the pyramidal neuron. Chandelier neurons target the pyramidal neuron axonic hilltop establishing multiple candle-like axo-axonic synaptic contacts (Chapter 6).

All of these functional features are applicable to the pyramidal neurons of all cortical strata. The combined interactions of excitatory and inhibitory synaptic contacts will determine the final function of the pyramidal neuron. Despite its importance and relevance, how a pyramidal neuron channel, organizes, and interacts with its numerous intrinsic and extrinsic functional inputs into a final motor action remain essentially unexplained.

It must also be pointed out, that a strict functional reciprocity exists between the pyramidal neuron ascending and stratified functional maturation and the development of its functional targets with proximal and distal neuronal systems, which may eventually involve millions of synaptic contacts. The timing of these functional reciprocities between a developing pyramidal neuron ascending maturation and its various neuronal (proximal and distant) targets remain practically unknown. During prenatal and postnatal development, these functional reciprocities will surely undergo progressive modulations and rearrangements. Consequently, an injected stem cell, even if it becomes a pyramidal neuron, will be incapable of reduplicating the innumerable developmental and functional reciprocities of the neuron that it intends to replace. Moreover, neurons interconnected with the damaged ones, unaffected by the original cortical insult, will have already rebuilt their synaptic profiles and assume new and different ones (Marín-Padilla 1996, 1997, 1999, 2000; Marín-Padilla et al. 2002, 2003). These acquired post-injury functional territories will be difficult if not impossible to be altered and/or replaced by the injected stem cell terminals. In my opinion, the idea that a transplanted stem cell might be capable of recovering the lost cerebral function is unrealistic. This idea ignores the brain extraordinary neuronal and fibrillar inter-complexity and its ongoing developmental and functional remodulations and readaptations.

In rapid Golgi preparations, it is not infrequent to observe contacts between an axon terminal and the dendritic spines (Figs. 4.5b, c and 4.6a–e). These contacts may be single and/or multiple (Figs. 4.5b, c). The multiple axo-spino-dendritic contacts, first described in 1968, are undoubtedly the most prominent and, functionally, the most significant (Marín-Padilla 1968). Multiple synaptic contacts on dendritic spine by a single axon terminal have been seldom described in the literature. The axon terminal runs parallel to the apical dendrite, for a considerable distance, and establishes visible contacts with the spines (Figs. 4.5b and 4.6a–e). They resemble a ladder in which the axon terminal and the dendrite represent its sides and the spine-contacts, the steps (Fig. 4.6b–c). The axon terminal is thinner than the dendritic shaft; its thickness is comparable to that of the spine pedicle. Some of these parallel synaptic contacts are notable: in some, the axon terminal establishes contacts with the spines, bypasses a blood capillary, and continues making additional spine-contacts (Fig. 4.6a, c, d). Possible sources for these multiple synaptic assemblages include the axon terminals of other pyramidal neurons as well as the axon terminals from afferent fibers. Multiple axo-spino-dendritic synaptic contacts have been also recognized in the motor cortex of infants with 21-trisomy (Fig. 4.6e). In Golgi preparations, is impossible to determine if the contacting axon terminal is from a descending and/or an ascending fiber. Probably, both ascending and descending axon terminals participate in these multiple synaptic contacts.

Dendritic spine abnormalities have been described in a variety of neurological disorders. Spine anomalies were first described in chromosomal trisomies (Marín-Padilla 1972, 1974, 1976) and have been subsequently in other mental disorders (Purpura 1974, 1975). In 13–15 trisomy, the dendritic spines are fine, long, and irregular and their number may be locally reduced (Fig. 4.7a). The long pedicles of the spine have focal enlargements mimicking synaptic contacts (Marín-Padilla 1974). In trisomy 21, the spines are also long and irregular with very fine, almost invisible, pedicles and prominent terminal heads (Marín-Padilla 1976). The fine spine pedicles are poorly stained because the amount of deposited silver in them is negligible, making them practically invisible (Fig. 4.7b). In the motor cortex of infants with 21-trisomy, some pyramidal neurons show signs of cystic degenerations affecting both their soma and apical dendrites (Fig. 4.7c, d). Pyramidal neurons with degenerating and spineless apical dendrites may be contiguous to unaffected ones with long, irregular spines with thin pedicles (Fig. 4.7c). The neuronal degeneration observed in these infants' motor cortex could explain the Alzheimer-like symptoms that often accompany this chromosomal disorder (Marín-Padilla 1976). The progressive cystic degeneration of these neurons remains unexplained. Perhaps, abnormal spines become functionally incompetent and could, eventually, be reabsorbed. The elimination of many abnormal spines will render the neuron incapable of functioning, resulting in its progressive degeneration. Recent studies describe additional features concerning the spine nature and functional and morphological features (Shepherd 1996; De Felipe 2006; Arrellano et al. 2007).

4.4 Ascending Functional Maturation

The mammalian pyramidal neuron is further characterized by two interrelated functional phases: an early developmental one characterized by its ascending sequential and stratified maturation (Chapter 3) and,

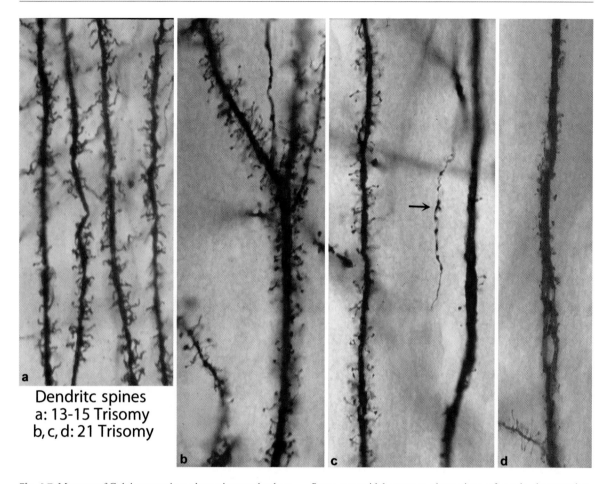

Fig. 4.7 Montage of Golgi preparations photomicrographs showing mid-segments of apical dendrites from P1 pyramidal neuron of the motor cortex of children with chromosomal anomalies, including 13–15 (**a**) and 21 (**b**–**d**) trisomies. In 13–15 trisomy (**a**), the spines are long and irregular and locally absent. In 21 trisomy (**b**), the spines are also long and irregular with very fine pedicles and prominent terminal head. The amount of silver deposited in the fine pedicles is minimal making them difficult to see (**b**). Some pyramidal neurons show signs of cystic degeneration, involving soma and dendrites, and severe lose of dendritic spines (**c**, **d**). Degenerating Pyramidal neurons with advanced degree of cystic degeneration may appear contiguous to less non-affected ones (**c**). Also some axon terminals show beading, suggesting a degenerative process. The neuronal degeneration observed in this infant motor cortex could explain the Alzheimer-like symptoms often described in this chromosomal disorder

a later one, characterized by its eventual descending function. Both functional phases may be applicable to all pyramidal neurons of the cerebral cortex. The pyramidal neuron dual functional phases reiterate this mammalian neuron uniqueness.

The pyramidal neuron's early developmental maturation is an ascending and stratified process that parallels the establishment of functional contacts with ascending afferent fiber terminals. These fibers ascend from the white matter, penetrate into the developing PCP from lower to upper strata and establish sequentially ascending contacts with the pyramidal neurons. By the 15th week of gestation afferent fiber terminals (of thalamic origin) have reached the PCP lowest region and start the maturation of its older and deeper pyramidal neurons (Fig. 4.8b). At this age, these pyramidal neurons develop short basal dendrites and a few proximal apical dendritic spines, establishing the first pyramidal cell P1 functional stratum in the motor cortex (Fig. 4.8b). At this time, the distal apical dendritic segment of these neurons is still spineless and functionally immature; actually, they have not even been extended. Subsequently, by the 20th week of gestation, the P1 pyramidal neuron apical dendrite distal segment continues to elongate upwardly and to develop more dendritic spines. At the same time, the pyramidal neurons above this stratum start to mature functionally paralleling the arrival of corticipetal fiber terminals at this level and

4.5 Descending Function

TABLE 1. Timing, Thickness and Differentiation of First Lamina, Subplate Zone and Pyramidal Cell Plate (PCP) Strata, of Human Motor Motor Prenatal Development

Age (wg)	Length (mm)	Layer I (µm)	Pyramidal cell plate (µm)	Functional maturation pyramidal cell plate	Subplate (µm)
7–8	15–22 CR	PPL	–	–	PPL
10–11	35–40 "	25–35	100–120	Undiff. PCP	30–50
15–16	80–100 CH	50–60	300–500	Layers VI, P1 Undiff. PCP	250–300
18–20	140–160 "	90–100	700–800	Layers VI, P1, P2 Undiff. PCP	Undetermined*
24–26	200–250 "	125–135	900–1000	Layers VI, P1, P2 P3 Undiff. PCP	"
28–30	280–320"	150–170	1,300–1,500	Layers VI, P1, P2 P3, P4 Undiff. PCP	"
38–40	420–460"	250–300	1,800–2,100	Layers VI, P1, P2 P3, P4, P5 Undiff. PCP	"

Key: wg = weeks gestation; mm = millimeters; µm = micrometers; CR = crown-rump; CH = Crown-heel; PPL = Primordial plexiform layer; PCP = Pyramidal Cell Plate; P1 to P5 = Ascending Pyramidal Cell Strata; * The Subplate Zone limits cannot be determine as it intermingles within the white matter zone. (Modified from Marín-Padilla and T. Marín-Padilla, 1982)

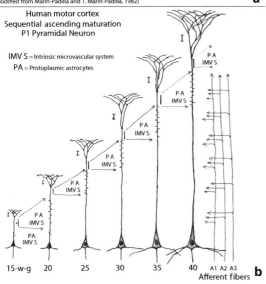

Fig. 4.8 Composite figure showing: the human motor cortex (**a**) progressive ascending developmental stratification including the PCP stratification between first lamina and subplate zone, and (**b**) a schematic representation of the sequential ascending functional maturation of P1 pyramidal neurons, of the motor cortex. These neurons ascending functional maturation also involve the anatomical and functional elongations of its basal, collaterals, and terminal dendritic bouquets. The functional maturation of pyramidal neurons continues during postnatal life (*arrow*). Other features of the composite figure are self-explanatory

establishing the motor cortex second pyramidal cell P2 functional stratum (Fig. 4.8b). During the subsequent functional maturation of the neocortex, the distal segment of the apical dendrites of P1 and P2 pyramidal neurons continue to elongate upwardly and to develop more dendritic spines. These events are repeated following an ascending progression paralleling the ascending penetration of terminals from thalamic, interhemispheric (callosal) and cortico-cortical fibers (Fig. 4.8b). Further maturation of all cortical strata pyramidal neurons will continue postnatally (Fig. 4.8b). The ascending human motor cortex and stratified functional maturation will establish sequentially the pyramidal cell P1, P2, P3, P4, P5, and P6 functional strata (Chapter 3). This type of ascending and sequential functional maturation is equally applicable to all pyramidal neurons of the PCP. Consequently, the pyramidal P6 stratum distinguishes the human motor cortex as human and different from that of other primates.

During the sequential ascending maturation of PCP pyramidal neurons the synaptic contacts include the terminals from thalamic, interhemispheric (callosal), and cortico-cortical fibers as well as those from intrinsic neuronal sources (Fig. 4.8b). The synaptic contacts from the various afferent fiber sources overlap along the length of the neuron apical dendrite (Marín-Padilla et al., 1969), such that pyramidal neurons from stratum P1 will receive inputs from all afferent fiber systems that reach the developing motor cortex. The number of synaptic contacts from thalamic fibers on intermediate pyramidal cell P3, P4 strata will decrease while those from callosal fiber will increase. The synaptic contacts on pyramidal neurons from the upper P5, P6 strata are primarily from cortico-cortical fibers, some from callosal fibers, and a few from thalamic terminals, such that pyramidal neurons from the upper strata are further distanced from the basic thalamic inputs to the cortex and closer to the more recent functional inputs (Fig. 4.8a).

The ascending and stratified functional maturation of pyramidal neurons represents a mammalian innovation that may be related to the cortex stratification and the number of cortical strata established during the course of mammalian phylogeny. The ascending and stratified gradient of the pyramidal neuron's synaptic profiles also play a crucial role in the number of strata established in the evolving mammalian neocortex. As evolving mammals' motor activities and capabilities increase, the number of supporting pyramidal cell strata in their motor also increases (Chapter 3).

4.5 Descending Function

All mammals share a pyramidal cell P1 stratum in their motor cortex that essentially will control the animal motor activity. The pyramidal neurons of P1 stratum function essentially control the animal motor activity. Their final function requires the functional participation of all pyramidal neurons from the above strata. Cajal, in his last book, published a year before his death, proposed a descending functional partway from upper to lower pyramidal neurons (Cajal 1933). This descending functional pathway interconnects all pyramidal neurons, from upper to lower strata, throughout the neocortex (Fig. 4.9b). The axon terminals of pyramidal neurons from the upper strata establish synaptic contacts with the dendrites of pyramidal neurons of the next lower strata, until all pyramidal neurons, from the different strata, become functionally interconnected (Fig. 4.9b). Such that P1 pyramidal neurons (the cortex main functional outlet) will receive input information from pyramidal neurons of all the above strata (Fig. 4.9b). I support and corroborate the idea of a descending functional partway interconnecting the pyramidal neurons of each stratum with those of lower strata, such that, in the mammalian motor cortex, the more recently added pyramidal cell stratum orchestrate the function of those of lower strata and eventually that of the essential pyramidal neurons of stratum P1. Recent neurophysiologic data seem to confirm an up-down functional pathway for the mammalian motor cortex excitatory network. "Thus, in motor cortex, descending excitation from a preamplifier-like network of upper-layer neurons drives output neurons in lower layers" (Weiler et al. 2008). These descending functional features are applicable to the pyramidal neurons of P2, P3, P4, P5, and P6 functional strata, respectively.

In addition, all pyramidal neurons' (from the various strata throughout the neocortex) terminal dendritic bouquets receive, within first lamina, a common type of input regardless of their eventual specific function, location, and/or cortical depth (Marín-Padilla and Marín-Padilla 1982; Marín-Padilla 1990). This common universal input originates in primordial corticipetal fibers that reach the cerebral cortex at the start of development and target the C-R cells dendrites. The C-R cells long horizontal axon terminals target the dendritic bouquets of all pyramidal neurons throughout the entire neocortex (Chapter 5). This universal input could represent a common physiologic tone, which will make all pyramidal cells, throughout the neocortex, capable of receiving and organizing more specific functional inputs, such that pyramidal neurons from primary, secondary, and tertiary motor, sensory, visual, and acoustic regions will all receive a common input from the first lamina as well as specific inputs from thalamic, interhemispheric, and cortico-cortical fibers and from specific intrinsic neuronal sources of the region (Chapters 5 and 6).

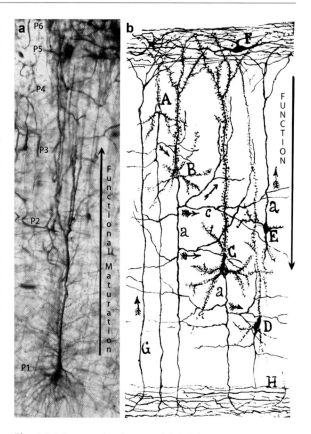

Fig. 4.9 Montage showing a rapid Golgi photomicrograph (**a**) of a newborn motor cortex depicting the ascending developmental maturation of pyramidal neurons and a reproduction (**b**) of an original Cajal drawing (Cajal 1933) illustrating his conception of pyramidal cells descending functional pathway. The axon terminals from upper strata pyramidal neurons are sequentially interconnected, by arrows, with the dendrites of pyramidal neurons of lower strata. While the functional maturation (**a**) of developing pyramidal neurons is an ascending and sequential process, their eventual function is a descending process (**b**), which sequentially interconnect pyramidal neurons from upper to those of lower strata. I agree with Cajal original conception of pyramidal neurons interconnecting and descending functional partway, which has been recently corroborated by neurophysiologic studies (Weiler et al. 2008)

In my opinion, the number of pyramidal cell strata formed in the developing motor cortex reflects the motor abilities and capabilities of each mammalian species. Any human, to accomplish all his/her motor needs and capabilities, requires an additional pyramidal cell P6 stratum that has more pyramidal neurons than the rest of the cerebral cortex (see Fig. 8.4). The sequential addition of pyramidal cell strata represents the base for the proposed new developmental cytoarchitectonics theory and nomenclature, which is applicable to all mammalian species (Chapter 9). In addition, the human newborn motor cortex (as well as

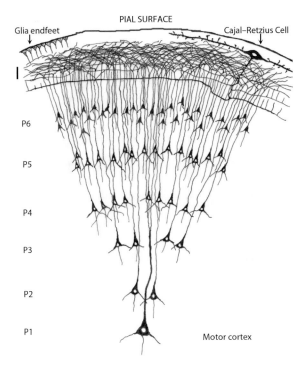

Fig. 4.10 Schematic representation of the functional territory of a single P1 pyramidal neuron of the motor cortex of a newborn infant. This neuron functional territory become defined by a descending functional partway contributed by all pyramidal neurons of the above strata (P2, P3, P4, P5, and P6). All terminal dendritic bouquets of pyramidal neurons are functionally interconnected, within the first lamina (I), by Cajal–Retzius (C-R) cell axonic processes. This scheme may be applicable to all mammalian species. The only differences will be in the number of pyramidal cell strata formed in their respective brains

that of other mammals) has a residual PCP stratum, beneath the first lamina, which will mature postnatally (see Figs. 2.14, 8.3, and 8.4). From developmental and evolutionary perspectives, the residual PCP postnatal functional maturation could contribute in the establishment of the additional pyramidal cell strata observed in the course of mammalian evolution (Chapter 9). In humans, it could participate in the incorporation of an additional P7 pyramidal cell stratum, which together with stratum P6, could represent the anatomical substratum for the higher motor and mental capabilities that distinguishes the human species (Chapter 9).

In conclusion, the function of a single pyramidal neuron of P1 stratum, of the motor cortex, is defined by the descending functional partway contributed by all pyramidal neurons of the above strata, which also defines the neuron functional territory (Fig. 4.10). The number of incorporated neurons increases from lower and older to superficial and younger pyramidal cell strata. To this very simplified model of the functional territory of a single pyramidal neuron, the inputs from corticipetal, callosal and cortico-cortical fibers systems as well as those from intrinsic non-pyramidal and of inhibitory neurons must be added. This concept should be applicable to the motor cortex of all mammalian species; the only difference will be in the number of pyramidal cell strata formed in their respective brains. Also, this simple interpretation of the cortex cytoarchitecture will help in explaining the bending of the brain surface, its progressive expansion, and the formation of cortical circumvolutions observed in the course of mammalian evolution.

The increase gyral pattern that characterizes the cetacean brain need to be better defined and explained. According to the observations presented herein, the cytoarchitectural stratification of the cetacean motor cortex should still be primitive and hence the number of pyramidal cell strata should be small despite the extraordinary surface expansion of their brain.

References

Arrellano JI, Benavides-Piccione R, De Felipe J, Yuste R (2007) Ultrastructure of dendritic spines: correlation between synaptic and spine morphology. Frontiers Neurosci 1:131–143

Cajal SR (1891) Sur la structure de l'ecorce cérébrale de quelques mammiferes. Cellule 7:124–176

Cajal SR (1894) La fine structure des centres des nerveoux. The Croonian Lecture. Proc R Soc Lond 55:443–468

Cajal SR (1896) Les spines collaterals des celulas du cerveux colorées au blue de méthylène. Revista Trimestrar Micrográfica 1:5–19

Cajal SR (1911) Histologie du systéme nerveux de l'homme et des vertebrés. Maloine, Paris (reprinted in Madrid, 1952)

Cajal SR (1923). Recuerdos de mi Vida. Juan Pueyo, Madrid

Cajal SR (1933) ¿Neuronismo o Reticularismo? Las pruebas objetivas de la unidad anatómica de las células nerviosas. Madrid (Reprinted by the Cajal Institute in 1952)

De Felipe J (2006) Brain plasticity and mental processes: Cajal again. Neurosciences 7:811–817

Fairen A, De Felipe J Regidor J (1984). Nompyramidal neurons: General Account, In Peters A, Jones GE (eds) Cerebral Cortex, Vol 1. Plenum Press, New York, pp 201–245

Gray EG (1959) Electron microscopy of synaptic contacts on dendritic spines of the cerebral cortex. Nature 183:1592–1593

Jones GE, Hendry SH (1984). The Basket Cell, In Peters A, Jones GE (eds) Cerebral Cortex, vol 1. Plenum Press, New York, pp 309–334

Marín-Padilla M (1967) Number and distribution of the apical dendritic spines of the layer V pyramidal cell in man. J Comp Neurol 131:475–490

Marín-Padilla M (1968) Cortical axo-spinodendritic synapses in man. Brain Res 8:196–200

Marín-Padilla M, Stibitz GR (1968) Distribution of the apical dendritic spines of the layer V pyramidal neurons of the hamster neocortex. Brain Res 11:580–592

Marín-Padilla M (1969) Origin of the pericellular baskets of the pyramidal cells of the human motor cortex. Brain Res 14:633–646

Marín-Padilla M, Stibitz GR, Almy CP, Brown HN (1969) Spine distribution of the layer V pyramidal neurons in man. A cortical model. Brain Res 12:493–496

Marín-Padilla M, Stibitz GR, Almy CP, Brown HN (1969). Spine distribution of layer V pyramidal cells in man. A cortical model. Brain Res 12:493–496

Marín-Padilla M (1970) Prenatal and early postnatal ontogenesis of the human motor cortex. A Golgi study, II. The basket-pyramidal system. Brain Res 23:185–191

Marín-Padilla M (1971). Early prenatal ontogenesis of the cerebral cortex (neocortex) of the cat (Felis domestica). A Golgi study. Part I. Z Anat Entzwickl-Gesch 134:117–145

Marín-Padilla M (1972) Structural abnormalities of the cerebral cortex in human chromosomal aberrations. A Golgi study. Brain Res 44:625–629

Marín-Padilla M (1974) Structural organization of the cerebral cortex (motor area) in human chromosomal aberrations. A Golgi study. I. D1 (13-15) trisomy. Patau syndrome. Brain Res 66:375–391

Marín-Padilla M, Stibitz JR (1974) Three-dimensional of the baskets of the human motor cortex. Brain Res 70: 511–514

Marín-Padilla M (1976) Pyramidal cell abnormalities in the motor cortex of a child with Down's syndrome. A Golgi study. J Comp Neurol 167:63–82

Marín-Padilla M, Marín-Padilla T (1982) Origin, prenatal development and structural organization of layer I of the human cerebral (motor) cortex. Anat Embryol 164:161–206

Marín-Padilla M (1987) The chandelier cell of the human visual cortex. J Comp Neurol 265:61–70

Marín-Padilla M (1990) The pyramidal cell and its local-circuit interneurons: a hypothetical unit of the mammalian cerebral cortex. J Cognitive Neurosci 2:89–105

Marín-Padilla M (1992) Ontogenesis of the pyramidal cell of the mammalian neocortex and developmental cytoarchitectonic: a unifying theory. J Comp Neurol 321:223–240

Marín-Padilla M (1998) The Cajal–Retzius cell and the development of the neocortex. Trends Neurosci (TINS) 21:64–71

Marín-Padilla M (1999) Developmental neuropathology and impact of perinatal brain damage. III: Gray matter lesions of the neocortex. J Neuropath Exp Neurol 58:407–429

Marín-Padilla M, Parisi JE, Armstrong DL, Sargent SK, Kaplan JA (2002). Shaken infant syndrome: Developmental neuropathology, progressive cortical dysplasia and epilepsy. Acta Neuropathologica 103:321–332

Marín-Padilla M, Tsai R, King MA, Roper SN (2003) Altered corticogenesis and neuronal morphology in irradiation-induced cortical dysplasia and epilepsy. A Golgi study. J Neuropath Exp Neurol 62:1129–1143

Purpura D (1974) Dendritic spines 'dysgenesis' and mental retardation. Science 186:1126–1128

Purpura DP (1975) Dendritic differentiation in human cerebral cortex: normal and aberrant developmental patterns. In: Kreutzberg GW (ed) Physiology and pathology of dendrites. Advances in neurology, vol 12. Raven, New York, pp 91–116

Peters A (1984) Bipolar cells. In: Peters A, Jones GE (eds) Cerebral cortex, vol 1. Plenum Press, New York, pp 381–405

Peters A, Saint Marie RL (1984) Smooth and sparsely spinous nonpyramidal cells forming local axonal plexuses. In: Peters A, Jones GE (eds) Cerebral cortex, vol 1. Plenum Press, New York, pp 419–442

Somogyi P, Cowey A (1984) Double bouquet cells. In: Peters A, Jones GE (eds) Cerebral cortex, vol 1. Plenum Press, New York, pp 337–358

Shepherd G (1996) The dendritic spine: a multifunctional integrative unit. J Neurophysiol 75:2197–2210

Valverde F (1967) Apical dendritic spines of the visual cortex and light deprivation in the mouse. Exp Brain Res 3:337–352

Weiler N, Wood L, Yu J, Solla SA, Shepherd GMG (2008) Top-down laminar organization of the excitatory network in motor cortex. Nat Neurosci 11:360–366

Human Motor Cortex First Lamina: Development and Cytoarchitecture

The first lamina is a primordial and fundamental organization of the mammalian cerebral cortex, which plays a crucial role on the development, placement, morphology, and ascending stratification of its essential neuron, the pyramidal cell (Marín-Padilla 1990). From the start of development, it represents an important functional compartment where the terminal dendritic bouquets of all pyramidal neurons, regardless of function (motor, sensory, visual, acoustic, or associative), size, and/or cortical depth, are progressively incorporated. Its basic cytoarchitectural organization remains essentially unchanged throughout the cortex prenatal and postnatal developments. Although its functional role is not well understood, it has been suggested that all pyramidal neurons throughout the cerebral cortex received, in this lamina, a common physiological tone, which activate and prepare them for the eventual acquisition of more specific functional roles (Marín-Padilla 1990; Marín-Padilla and T. Marín-Padilla 1982).

In formalin-fixed postmortem brain preparations stained with routine and universally used procedures (such as H&E and Nills), the first lamina appears as an essentially empty subpial stratum with a few scattered cells. This apparent emptiness has contributed to misconceptions about this lamina nature, composition, cytoarchitectural organization, and functional importance. In most histological and pathological descriptions as well as in most illustrations of the cerebral cortex, the first lamina is portrayed as a barren compartment with scattered neurons (see Fig. 3.14B). However, using appropriate staining procedures, its cytoarchitectural organization and composition are extraordinarily complex with several specific components and the terminal dendrites of all underlying pyramidal neurons. Moreover, its basic composition and cytoarchitectural organization remain essentially unchanged from the start of cortical development and through its subsequent prenatal and postnatal maturations (Marín-Padilla 1990; Martin et al. 1999).

During early cortical development, these lamina components include: primordial corticipetal fibers of unknown origin, large neurons with long horizontal processes recognized as Cajal–Retzius (C-R) cells, the terminal dendritic bouquets of all underlying pyramidal-like neurons of the subplate zone (SP), and the axonic terminal of SP Martinotti cells (Chapters 1, and 2). During subsequent development, the terminal dendritic bouquets of all underlying pyramidal neurons become progressively incorporated into it. To adequately visualize this lamina cytoarchitectural organization and composition it is necessary to use a procedure capable of staining all its elements. The rapid Golgi procedure remains the best available procedure capable of staining all its elements as well as their evolving organization and developmental interrelationships (Chapter 12).

The first lamina development, composition, cytoarchitecture, and functional organization described herein are essentially based on rapid Golgi studies of the developing human cerebral motor region. It is the first stratum recognized in the developing mammalian cerebral cortex. Its development starts with the early arrival, at the cerebrum subpial region, of primordial corticipetal fibers and the incorporation of a few neurons. Both of these components arrive from extracortical sources and extend throughout the cerebrum subpial zone in a proximal (ventral) to distal (dorsal) gradient (Chapters 2 and 3). The presence of primordial corticipetal fibers and of a few scattered neurons are first recognized in the cerebrum subpial zone in 6- and 7-week-old human embryos (Chapter 3). Using neurofilament staining procedures and Bodian silver stain, the internal capsule fibers can be followed up to the cerebral vesicle, observe their distribution and final penetration into the cerebrum subpial zone (Chapter 3). Their presence coincides with the recognition, throughout the subpial zone, of a few scattered neurons, interspersed among the fibers and with the expression of "reelin." This early composition (marginal zone stage) of the mammalian cerebral cortex is considered to be a primordial functional system that shares some features with the amphibian cortex (Marín-Padilla 1971, 1972, 1978, 1983, 1998). Subsequently, its neurons become segregated into superficial and deep ones. The superficial neurons, sandwiched among horizontal fibers, assume a horizontal morphology compatible to that of C-R cells and the deep ones assume a stellate morphology with ascending apical dendrites compatible with subplate (SP) zone pyramidal-like neurons. The primordial corticipetal fibers extend throughout the first lamina and functionally target the dendrites of both the C-R cells and the SP pyramidal-like neurons. Also, this early primordial plexiform (PP) cortical organization, already recognized in 7-week-old human embryos, represents a functional system, shares some features with the reptilian cortex, and is a prerequisite for the subsequent development of the cortex pyramidal cell plate (PCP) (Chapters 2 and 3).

Subsequently, ascending neuroblasts, generated at the ependymal epithelium, attracted by "reelin" and guided by radial glial fibers, start to accumulate within the PP forming the neocortex PCP and establishing, simultaneously, the cortex first lamina and the SP zone, above and below it, respectively. The first lamina C-R cells would play a crucial role in the formation, ascending functional stratification, and placement of all pyramidal neurons of the mammalian cerebral cortex. The PCP represents a mammalian innovation and the most distinguishing feature of the neocortex (Chapters 2 and 3). In humans, the PCP formation starts around the 8th week of gestation and is nearly completed by the 15th week of gestation (Chapter 3).

From the start of development, the first lamina is characterized by the organization of three essential components, namely: the C-R cells, the primordial corticipetal fibers, and the terminal dendrites of the pyramidal neuron. These three components are anatomically and functionally interconnected and share a common horizontal functional field that extends throughout the entire surface of the cortex. This shared and common horizontal functional field expands during cortical development as the dendritic bouquets of new pyramidal neurons become progressively incorporated into the lamina (Figs. 5.1–5.4).

Various secondary components (small local-circuit neurons, some terminals from thalamic, callosal, and cortico-cortical fibers and the axon terminals of Martinotti cells) are later incorporated and will not affect the lamina basic and universal horizontal cytoarchitectural organization. The first lamina functional horizontal field extends through the entire neocortex and is composed of the long horizontal terminals primordial corticipetal fibers (afferent system) that functionally target the C-R cell dendrites (receptive system) and the C-R cells long horizontal axon terminals that target the dendritic bouquets of all pyramidal neurons (efferent system). The primordial corticipetal fibers also target the dendritic bouquets of pyramidal neurons. There are no reasons to think that this primordial functional system will cease to function during the cortex postnatal development. Neither the C-R cells, the primordial corticipetal fibers (although both significantly diluted), or the pyramidal neurons terminal dendritic bouquets of pyramidal neurons (significantly augmented) disappear from the cortex. Later in development, the dendritic bouquets of pyramidal neurons belonging to the different cortical regions will also receive (locally) additional inputs from some thalamic, interhemispheric (callosal), and corticocortical afferent fiber terminals as well as from local-circuit neurons, which will determine their specific functions.

During early developmental stages (from the 7th to the 15th week gestation), the first lamina has also the terminal dendritic bouquets of the SP zone pyramidal-like neurons. During early developmental stages, these terminal dendritic bouquets of neurons represent the first lamina most prominent elements and are essential components of the mammalian neocortex early primordial functional organization. By the 15th-week gestation, the terminal dendritic bouquets of neurons start to lose their first lamina functional contacts and, subsequently, start to regress (Chapter 3). At this embryonic stage, the neocortex early primordial functional organization based on the SP zone pyramidal-like neurons in conjunction with C-R cells start to be replaced by its definitive function system based on its pyramidal neurons also in conjunction with the C-R cells (Chapter 3).

5.1 First Lamina Principal Components

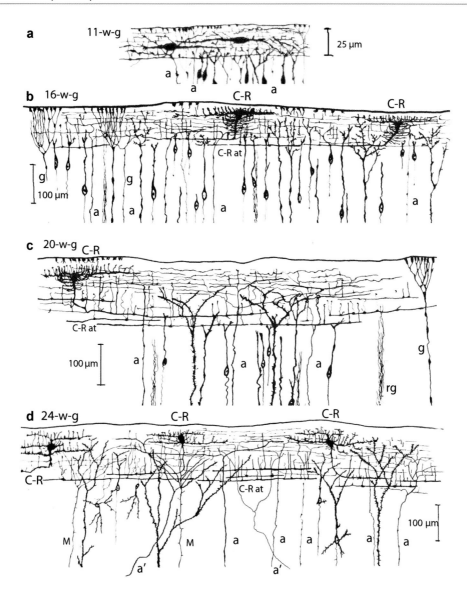

Fig. 5.1 Montage of camera lucida drawings, from rapid Golgi preparations, showing the development of the human motor cortex first lamina cytoarchitectural organization and composition through the 11th, 16th, 20th, and 24th week of gestations. The first lamina essential neuronal and fiber composition changes very little during prenatal development. Early in development (**A**), the Cajal–Retzius (C-R) cells are large horizontal bipolar neurons sandwiched among the lamina numerous horizontal fiber terminals. Later in development (**B**, **C**), the C-R cells assume embryonic features characterized by the concentration of their dendritic and axonic processes into a reduced space and by their displacement toward the lamina upper region. By the 24-week gestation, C-R cells start to lengthen their processes, paralleling the cortex expansion, and start to assume more mature morphological features. The cortex first lamina is also characterized by numerous horizontal fiber terminals, which expand its entire thickness, representing both C-R cell horizontal axonic collaterals and terminal axon and horizontal terminals from primordial corticipetal fibers. By the 20th–24th week of gestation (**C**, **D**) the lamina horizontal fiber terminals are segregated into those coursing through its upper region and those through its lower one. While C-R cells horizontal axon terminals tend to run through the lamina lower region, corticipetal fiber terminals run through its upper region coinciding with the neuron dendritic processes. Other first lamina components include the terminal dendritic bouquets of pyramidal neurons and the radial glia terminal filaments. Key: C-R at, C-R axon terminal; G gl, glia; rg, radial glial fibers; a, afferent fibers; M, Martinotti axon terminals. Scales: 25 (**A**) and 100 μm (**B**, **C**, **D**)

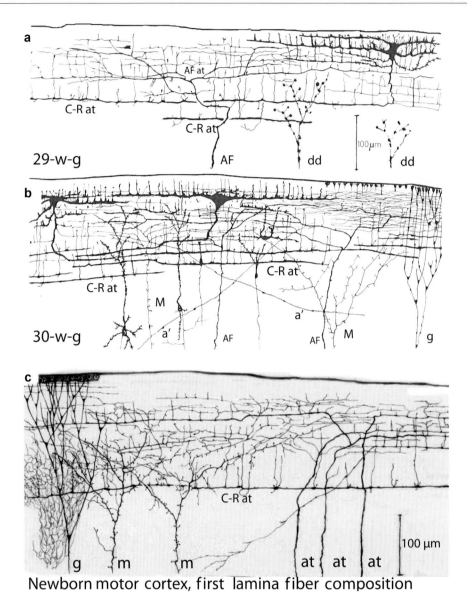

Newborn motor cortex, first lamina fiber composition

Fig. 5.2 Montage of camera lucida drawings, from rapid Golgi preparations, showing the human motor cortex first lamina cytoarchitectural and organization of fetuses of 29- (**A**) and 30-week of gestation (**B**) and of a newborn (**C**). The developmental lengthening "horizontalization" of C-R cells dendritic and axon terminals and primordial corticipetal fiber terminals has progressed, paralleling the cortex expansion. Corticipetal fiber reached the lamina upper region and branch into long horizontal terminals coinciding with the distribution of C-R cell dendrites. The C-R cells descending axons give off numerous horizontal collaterals distributed throughout the lamina middle region and become long horizontal terminal (C-R at) throughout its lower region. Also illustrated are Martinotti cell (M) terminal axonic bouquets (**B**, **C**), some degenerating (dd) terminal dendrites (**A**). Scales: 100μm

The neocortex first lamina has also nonfunctional components, including innumerable radial glial filaments with terminal endfeet that, united by tight junctions, build, maintain, and repair the neocortex external glial limiting membrane (EGLM) and manufacture its basal lamina material (Figs. 5.1B, C and 5.2B, C). The EGLM demarcates the neocortex (as well as the entire central nervous system) from surrounding meningeal tissues, maintains its anatomical and functional integrities and must be preserved intact throughout both prenatal and postnatal developments. Any EGLM rupture (mostly caused by pathological conditions) has to

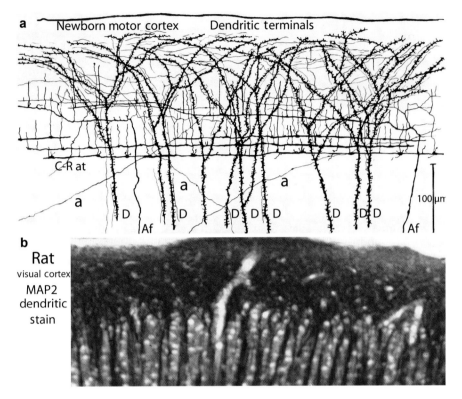

Fig. 5.3 Composite figure with a camera lucida drawing (**A**) and a photomicrograph from a MAP2 dendritic stained preparation (**B**) showing the large concentration of terminal dendritic bouquets of pyramidal neurons that characterized the mammalian cerebral cortex first lamina. The pyramidal terminal dendrites represent the first lamina principal receptive system, which is functionally targeted by both corticipetal fiber terminals and C-R cell axon terminals. The light tubular structures seen in the middle of Fig. **B** represents a perforating capillary

be repaired. The EGLM reparation often resulted in the formation of leptomeningeal heterotopias, frequently associated with epilepsy (Marín-Padilla 1996, 1997, 1999, 2002). During prenatal development, the number of radial glia fibers increases progressively supplying the necessary additional glial endfeet for the EGLM of the expanding cortical surface. However, during later developmental stages, a special type of astrocytes becomes progressively incorporated into the first lamina contributing additional endfeet to the cortex expanding EGLM. Eventually, these first lamina special astrocytes will replace the radial glia as the main source of endfeet for the EGLM of the expanding neocortex (Chapter 8).

During late prenatal development, a variety of small local-circuit neurons are also incorporated into the first lamina. The local-circuit interneurons functional target is also the terminal dendritic bouquet of pyramidal neurons However, their anatomical and functional fields are restricted and localized and may vary from region to region. Their local and restricted functional fields on the terminal dendrites of pyramidal neurons contrast sharply with the universal and shared functional field established by C-R cells long horizontal axonic terminals. The appearance, cytoarchitecture, and possible functional contributions of these late-incorporated secondary elements are explored below.

Eventually, growing capillaries from the pial anastomotic capillary plexus perforate the neocortex EGLM, enter into the neocortex, and progressively establish its extrinsic and intrinsic microvascular systems (Chapter 7).

5.1 First Lamina Principal Components

From the start of development, C-R cells are recognized as large neurons sandwiched among corticipetal fibers with long horizontal processes, which tend to

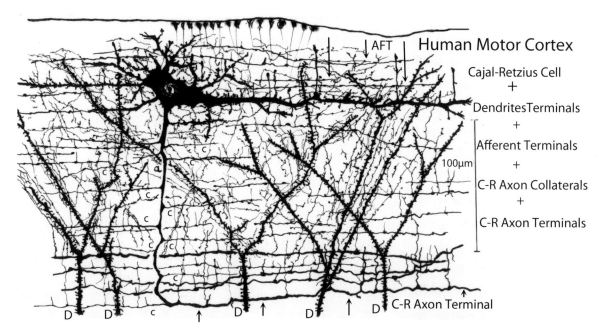

Fig. 5.4 Detailed camera lucida drawing, from Golgi preparations, illustrating, at a higher magnification, the motor cortex first lamina essential functional system and its components. Afferent fiber terminals (AFT) reach and branch, throughout the lamina upper stratum, into long horizontal collaterals that coincide with C-R cells dendrites distribution and the terminal segments of dendrites of the pyramidal neurons. The C-R cell descending axon give off numerous horizontal collaterals (c) distributed throughout the lamina middle stratum coinciding with the distribution of terminal dendrites of pyramidal cells. The C-R cell axon reaches the lamina lower stratum and becomes a long horizontal fiber (tangential fibers of Retzius). All the horizontal axonic terminals of corticipetal fibers, of C-R cells and the C-R cell terminal axons have numerous ascending and fewer descending branches distributed among the terminal dendrites of pyramidal cells, suggesting both morphological as well as functional interrelationships

occupy the lamina upper zone beneath the pial surface (Chapters 2 and 3). The first Golgi preparations of the human motor cortex available are from an 11-week-old fetus (Fig. 5.1A). At this stage, C-R cells are bipolar neurons with prominent horizontal dendrites and axonic terminals. No other neuronal types are recognized in the lamina at this stage. Other first lamina components include the long horizontal axonic terminals of primordial corticipetal fibers, the terminal dendritic bouquets of SP zone pyramidal-like neurons, and those of newly incorporated pyramidal neurons (Fig. 5.1A). By the 15th week of gestation, the C-R cells have already assumed their distinctive morphological appearance (Fig. 5.1B). They are large bipolar neurons with horizontal dendrites with numerous short ascending branches and a descending axon with horizontal collaterals that reaches the lower strata and becomes a long horizontal axonic terminal (Fig. 5.1B). The concentration of dendritic processes closer to the neuron soma corresponds to its early or embryonic morphology. The neuron has not yet started its progressive horizontal expansion and/or lengthening of its dendritic and axonic processes (Fig. 5.1B). The progressive lengthening of C-R cells processes is referred herein as: developmental "horizontalization."

By the 20th week of gestation, the neocortex has expanded by the continued incorporation of additional dendritic bouquets of pyramidal neurons; consequently the C-R cells have started the progressive horizontalization of their dendritic and axonic processes (Fig. 5.1C). By the 24th week of gestation, the C-R cell dendritic and axonic processes horizontalization is at an intermediate stage and are recognized throughout the entire neocortex first lamina (Fig. 5.1D). However, the recognition of the C-R cell neuron body, within a single Golgi preparation, decreases progressively. The actual number of C-R cells already established at the start of neocortical development will not increase further. Therefore, as the neocortex expands, these neurons undergo a progressive and significant developmental dilution. However, their long horizontal axonic processes are recognized throughout the neocortex. It is assumed that the C-R cell body and main dendrites retain their original location and functional contacts with the primordial corticipetal fibers while the neocortex expands progressively. Consequently, the localization of C-R cells and of primordial corticipetal fiber become increasingly difficult as the cortex expands. The assumed developmental disappearance of C-R cells, a generally held idea, is in my opinion mistaken.

The early primordial corticipetal fibers are recognized because they ascend from the white matter, cross unbranched the PCP without making contacts with its pyramidal neurons, reach the first lamina, and become long horizontal axonic terminals (Figs. 5.1–5.4). While these long horizontal fibers are recognized throughout the neocortex first lamina, detecting them, in single rapid Golgi preparations, as they ascend through the PCP also becomes increasingly difficult. Their original number, as that of C-R cells, was established at the start of neocortical development and will not increase further. As the neocortex expands the possibility of detecting an ascending primordial corticipetal fiber as it crosses the PCP decreases accordingly.

By the 24th week of gestation, the first lamina long horizontal fiber terminals are starting to be segregated into upper and lower systems (Fig. 5.1D). Fibers distributed through the lamina upper zone are thinner and mostly represent primordial corticipetal fibers terminals. Their superficial distribution coincides with that of C-R cell dendrites (Figs. 5.1–5.4). The horizontal fibers coursing throughout the first lamina lower zone are thicker and mostly represent the C-R cells long horizontal axonic terminals (Figs. 5.1C, D, 5.2A–C, and 5.3A). The descending C-R cells axons giving-off several horizontal collaterals distributed through the lamina middle stratum, which coincide with the dendritic distribution of pyramidal neurons (Figs. 5.1D, 5.2A–C, 5.3A). The first lamina early fiber stratification will be maintained essentially unchanged throughout the neocortex prenatal development (Figs. 5.1C, D, 5.2A–C, 5.3A, and 5.4). The continued incorporation and the local anatomical and functional expansions of the terminal dendrites of the pyramidal neurons result in the progressive lengthening (horizontalization) of C-R cell and primordial corticipetal axonic processes. The concentration of terminal dendrites of the pyramidal neurons within the first lamina becomes the larger of the developing neocortex. Moreover, dendritic stains (MAP2) demonstrate that the concentration of dendrites within the first lamina continues to be the larger even in the adult brain (Fig. 5.3B). The first lamina stratified cytoarchitectural organization as well as the structure and functional interrelationships of its basic components are best illustrated using camera lucida montages, obtained from rapid Golgi preparations (Figs. 5.1–5.4).

During neocortical development, while the first lamina thickness and horizontal expansion increases progressively by the incorporation of additional terminal dendrites of the pyramidal neurons, the overall cytoarchitectural organization of its basic components remains essentially unchanged (Figs. 5.1–5.8).

5.2 Cajal–Retzius Cell Unique Morphology

The rapid Golgi procedure remains the best, possibly the only, staining method capable of staining the C-R cells as well as their developmental cytoarchitecture, distribution and functional interrelationships (Figs. 5.1–5.12). In Golgi preparations cut perpendicular (vertical) to the precentral gyrus long axis (the most common type of preparation used), C-R cells assume, at least, three different morphologic appearances. The neuron may appear as a bipolar cell with prominent horizontal dendrites emerging from the soma (Figs. 5.5A and 5.6B), as a monopolar neuron with a single horizontal dendrite (Figs. 5.5D and 5.6C), and/or pear (pyriform)-shaped neuron with short and thin dendrites (Figs. 5.5B, C and 5.6A, D). The original Golgi descriptions of this neuron by Cajal, Retzius, and myself illustrate its triple morphological appearance (Cajal 1891, 1911; Retzius 1893, 1894; Marín-Padilla and T. Marín-Padilla 1982; Marín-Padilla 1990, 1992). In contrast, the neuron descending axon, its distribution, collaterals, and long horizontal axonic terminal are always similar in the three morphologic types (Figs. 5.1D, 5.2B, 5.4A, 5.5A–D, and 5.6A–D). Invariably, the neuron descending axon emerges from the soma, crosses the lamina middle stratum giving-off several horizontal collaterals, and becomes a thick long horizontal terminal running through its lower zone (Figs. 5.4A, 5.5A–D, and 5.6A–D). The similarity of these neurons axonic features are also described and illustrated in Retzius, Cajal, and my original Golgi studies (Figs. 5.1–5.6). The neuron long axonic

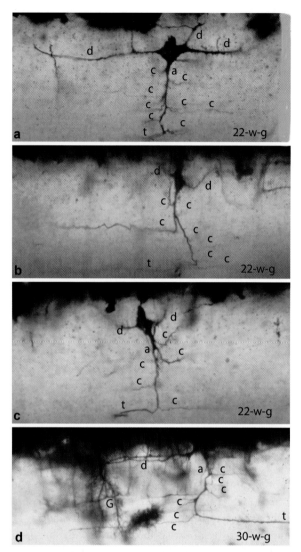

Fig. 5.5 Montage of photomicrographs, from rapid Golgi preparations, of the motor cortex of fetuses, illustrating the contrast between the C-R cells variable dendritic morphology and the similarity of their descending axons and collaterals. C-R cells may appear as bipolar neurons with two prominent horizontal dendrites (**A**), as pyriform with very short dendrites (**B, C**) and/or as monopolar with a single horizontal dendrite (**D**). In contrast, their axon is invariably a descending fiber that give off several horizontal collaterals (c) through the lamina middle region and becomes a long horizontal terminal (t) throughout its lower region (**A, B, C, D**)

processes are further characterized by the presence of numerous ascending short branches distributed among the terminal dendrites of pyramidal neurons (Fig. 5.4). Moreover, the C-R cells long horizontal axonic terminals are a prominent feature of the first lamina recognized throughout the entire neocortex (Figs. 5.1–5.6).

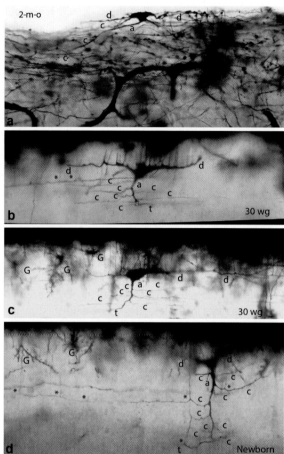

Fig. 5.6 Montage of photomicrographs, from rapid Golgi preparations, of the motor cortex of 2-month-old infant (**A**), 30-week-old fetuses (**B, C**), and of a newborn (**D**) illustrating the persistent disparity between the C-R cells variable dendritic morphology and the uniformity of their axonic features. The C-R cell soma (**A**) may appear irregular with thin dendrites (d), as bipolar (**B**) with two prominent horizontal dendrites (d), as monopolar (**C**) with a single dendrite (d) and/or with an irregular pear shaped soma (**D**). Contrarily, the axon of these neurons is invariably a descending fiber that gives off several collaterals (c) and becomes a terminal (t) axonic fiber coursing through the lamina lower region. Also illustrated are several first lamina astrocytes (G) and some of the ascending short branches (*) of C-R cell axonic collaterals

Moreover, these long and thick horizontal axonic terminals (tangential fibers of Retzius) are among the first elements of the neocortex to myelinate, corroborating their functional importance (Rocker and Riggs 1969; Yakovlev and Lecours 1967). Myelinated horizontal fibers running through the first lamina lower zone are recognized in the motor cortex of both newborns and adults brains (Marín-Padilla, 1990).

The C-R cells of the human cerebral cortex, despite a few and nearly identical morphologic descriptions (Cajal, Retzius, and myself), have remained enigmatic and controversial and have been seldom described in the literature (Marín-Padilla and T. Marín-Padilla 1982; Marín-Padilla 1990). While the neuron long axonic processes are recognized throughout the entire neocortex, the location of its body has remained elusive and difficult to locate. This difficulty has been, erroneously, interpreted as the neuron developmental disappearance. The progressive and continued incorporation of terminal dendritic bouquets from pyramidal neurons (the C-R cell neuron main functional targets) or the persistence, throughout the neocortex, of the neuron long horizontal axonic processes do not support its developmental disappearance.

A dilemma that needs to be resolved is the disparity between the neuron triple dendritic morphologic and the parity of its axonic features and distribution. Searching for explanations of this dilemma, Golgi preparations were cut both perpendicular and parallel to the long axis of the motor precentral gyrus (Figs. 5.7A–C and 5.8A–D). It turns out that the first lamina fiber composition, stratification, and distribution remain unchanged and indistinguishable in Golgi preparation cut parallel (Fig. 5.7A, C) and/or perpendicular (Fig. 5.7B, D) to the precentral gyrus long axis. Similarly, the C-R cell triple dendritic morphology continues to be recognized in section cut perpendicular and/or parallel to the gyrus long axis. Therefore, the neuron dendritic and axonic morphology, stratification, and distribution are essentially identical and impossible to distinguish in perpendicular and/or parallel cut Golgi preparations of the precentral gyrus. As expected, the profiles of terminal dendrites of the pyramidal neurons also remain unchanged and indistinguishable in perpendicular and/or parallel cut Golgi preparations. The progressive and universal incorporation of terminal dendrites of pyramidal neurons throughout the first lamina will maintain their profiles unchanged from any angle of view.

Since, neither perpendicular nor parallel cut Golgi preparations of the motor cortex solve this C-R cell dilemma, additional tangentially cut Golgi preparations of the precentral gyrus were made. To visualize tangentially the first lamina cytoarchitectural organization without interferences, new Golgi preparations were made reducing the fixation and especially the silver impregnation times to a minimum of only a few hours (Chapter 12). By reducing the silver impregnation time it was possible to view tangentially the entire first lamina thickness without interferences and/or silver precipitates.

Since the thickness of a Golgi preparation ranges from 150 to 250 μm, a good tangential section should provide independent views of the lamina upper, middle, and lower strata (Figs. 5.8–5.12). Tangential views of first lamina upper stratum, immediately beneath the pial surface, demonstrate the presence of the C-R cell soma and its main dendrites, some special astrocytes and numerous fine fibers, coursing in all directions, representing primordial corticipetal fiber axonic terminals (Fig. 5.8A). Although, hundreds of tangential sections may be required to view a single C-R cell tangential views through the lamina middle stratum show numerous fine horizontal fibers, coursing in all directions, representing C-R cell axonic collaterals, some axonic terminals from corticipetal fibers, and pyramidal neurons terminal dendrites tangential cuts (Fig. 5.8B). Tangential view through the lamina lower stratum show numerous thick horizontal fiber terminals, also coursing in all directions, representing the C-R cells horizontal axonic fibers (Fig. 5.8C). A multidirectional expansion of all horizontal axonic processes, within the first lamina, is to be expected because they respond and adjust to the cortex multidirectional expansion, which result from the universal incorporation of terminal dendrites of pyramidal neurons. Consequently, during cortical development, the horizontal axonic processes C-R cells and of corticipetal fibers must expand in all directions to maintain functional contacts with the increasing number of terminal dendrites of pyramidal neurons progressively incorporating into first lamina. From a developmental perspective, it is not possible to separate the C-R cell from its functional targets and is equally impossible to support the idea that such a fundamental neuron should disappear from the cortex when its functional targets are not.

Moreover, in tangential Golgi preparations, all C-R cells observed appear as multipolar neurons with horizontal (tangential to pial surface) dendrites that also branch, out of its body, in all directions (Figs. 5.8A, 5.9A–E, 5.10A, and 5.12). Consequently, in vertically cut Golgi preparations, either parallel and/or perpendicular to the precentral gyrus long axis, the C-R cell dendritic morphology could indeed appear as bipolar, monopolar, and/or pear shaped, depending on the angle of the cut section. Nearly a hundred years after Cajal's

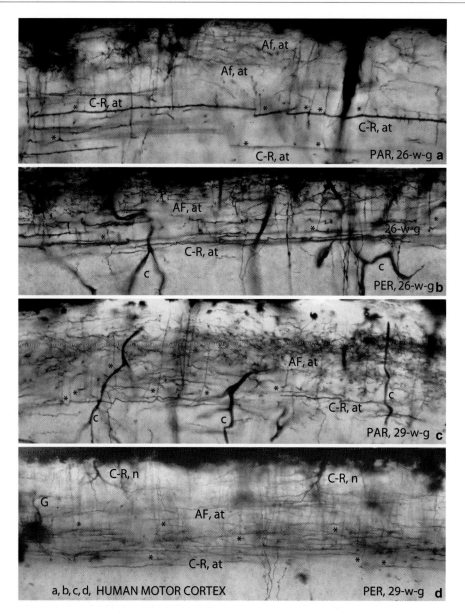

Fig. 5.7 Montage of photomicrographs, from rapid Golgi preparations, of the motor cortex of 26- and 29-week-gestation fetuses, showing that first lamina fiber components distribution and stratification remain essentially unchanged sections cut parallel (**A**, **C**) and/or perpendicular (**B**, **D**) to the long axis of the precentral gyrus. The finer horizontal fibers from afferent corticipetal fibers (Af, at) tend to occupy the lamina upper region while the thicker ones, representing C-R cells axon terminals (C-R, at) are found through its lower region. Many ascending terminal branches (*) from C-R cells horizontal axon terminals are recognized in parallel (**A**, **C**) as well as in perpendicular (**B**, **D**) cut preparations. Some capillaries (c) from the first lamina intrinsic microvascular system are recognized in all preparation. Two C-R cells (C-R n) are also recognized in preparation **D** (see also Figs. 5.1, 5.2, and 5.3A)

original description, this neuron multipolar dendritic morphology was finally established using tangential Golgi preparations (Marín-Padilla 1990).

It must be pointed out, that both Cajal and Retzius, who also made first lamina tangential preparations, could have also recognized this neuron multipolar morphology (Retzius 1894; Cajal 1891). Retzius used tangential Golgi preparations of fetuses and newborn infant's cerebral cortex and Cajal the Ehrlich méthyléne blue method of adult cat's cerebral cortex,

5.2 Cajal–Retzius Cell Unique Morphology

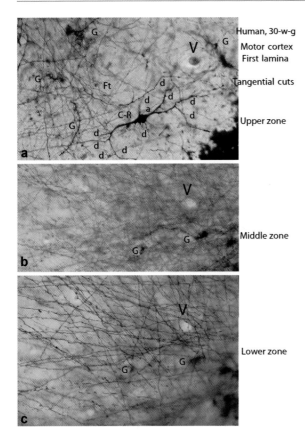

Fig. 5.8 Montage of photomicrographs, from rapid Golgi preparations, of a 30-week-gestation fetus showing three – upper (**A**), middle (**B**), and lower (**C**) – tangential sequential views of first lamina cytoarchitectural organization. The same perforating vessel (V) is recognized through the three levels. The upper level (**A**) shows a tangential view of a C-R cell soma with multipolar dendrites (d) and emerging axon (a), numerous fine tangential fibers coursing (crisscrossing) in all directions representing corticipetal fibers axonic terminals (Ft) and a few special astrocytes (G). The middle level (**B**) shows innumerable very fine tangential fibers also coursing (crisscrossing) in all directions, which represent C-R cells axonic collaterals as well as two special first lamina astrocytes. The lower level (**C**) shows numerous thick tangential fibers coursing (crisscrossing) in all directions representing C-R cells axon terminals and part of the two astrocytes seen on **B**. Remnants of silver deposits are also recognized in the left lower corner of all photos

respectively. It is quite possible that both recognized this neuron multipolar dendritic morphology, which later since Kölliker work (1896) came to be known as Cajal–Retzius cells. Laminas XIV, XVI, XVII, and XIX from Retzius work and Fig. 2 from Cajal's illustrate the multipolar dendritic morphology of first lamina large neurons. In my opinion, Retzius and Cajal, and even Kölliker, fail to recognize that these large multipolar neurons actually represent different views of the neuron they have described in vertically cut Golgi preparations. Consequently, in Retzius and Cajal's original works (and even in my early one), only monopolar, bipolar, and/or pyriform dendritic patterns were described for these large first lamina neurons (Cajal 1891; Retzius 1894; Marín-Padilla and T. Marín-Padilla 1982).

Tangential Golgi preparations of the human motor cortex have also demonstrated that the C-R cell dendritic morphology is quite variable (Figs. 5.8A–C, 5.9A–E, 5.10A, 5.11A, B, and 5.12). Their dendritic processes have variable length and thickness and expand horizontally (tangential to pial surface) in different direction (Fig. 5.9A–E). They also have numerous smaller ascending and descending branches. Their descending axon gives off several collaterals that branch horizontally and in many directions. Finally, its descending axon reaches the lamina deep stratum and becomes a single thick horizontal axonic terminal that expands in a single direction (Figs. 5.9B, 5.10B, and 5.12).

A montage of camera lucida drawings, from tangentially cut Golgi preparations of the motor cortex of 30-week-old fetuses; illustrate, more clearly, these neurons actual dendritic (Fig. 5.10A) and axonic tangential features (Fig. 5.10B). During development, a C-R cell expands in all direction (concentrically) its axonic terminals covering progressively larger circular functional territories. This centrifugal axonic expansion responds to the universal incorporation of dendritic terminals of pyramidal neurons throughout the cortex first lamina. Eventually, the tangential and circular functional territories of contiguous C-R cells (although progressively distanced) will overlap, with each other, eventually covering the entire neocortex and contacting all its pyramidal neurons terminal dendrites. Consequently, while the expanding functional territories of C-R cells may be recognized throughout the entire cortex, the location of a single neuronal body becomes increasingly difficult to find. Hundreds (even thousands) of preparations will be required to, perhaps, locate a single neuronal body. The number of slides needed to find a single C-R cell increases exponentially as the cortex progressively expands. Searching through thousands of H&E preparations, I have been able of localizing only but a few C-R cell bodies (Marín-Padilla 1990; Martin, Gutiérrez Peñafiel and de la Calle 1999). On the other hand, myelin stained (Kluver-Barrera stain) preparations of the adult human motor cortex, cut perpendicular and

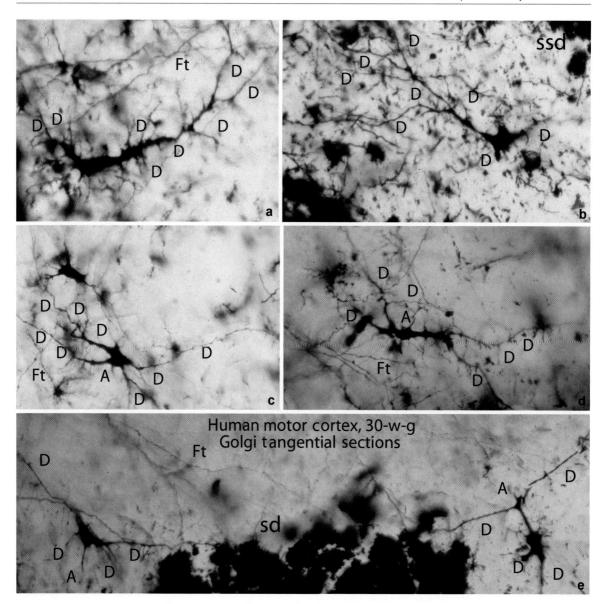

Fig. 5.9 Montage of photomicrographs, from rapid Golgi preparations, of the motor cortex of a 30-week-gestation human fetus showing tangential views (**A, B, C, D, E**) of first lamina upper level C-R cells with variable multipolar dendritic morphologies. Among these neurons tangential dendrites (D) and emerging descending axons (A) are fine tangential terminals (Ft) from corticipetal fibers. Small silver deposits are found in some photos and a large one (sd) covers part of Fig. **D** (see also Fig. 5.10A)

parallel to gyrus axis, demonstrate the presence of horizontal thick myelinated fibers coursing through its middle and lower strata. Moreover, in myelin stained tangential preparations, the same myelinated fibers are recognized to crisscross the first lamina lower strata in all directions (Marín-Padilla, 1990). These myelinated fibers coursing through the first lamina lower zone can only represent the C-R cells long horizontal axonic terminals. In my opinion, the C-R cell nature and fundamental functional target explain, by themselves, its essential morphological and functional features. It is also important to point out, that the neuron axonic collaterals and terminal axonic fibers have innumerable ascending and some descending processes that branch in close proximity to the terminal dendrites of the pyramidal neurons (Fig. 5.4).

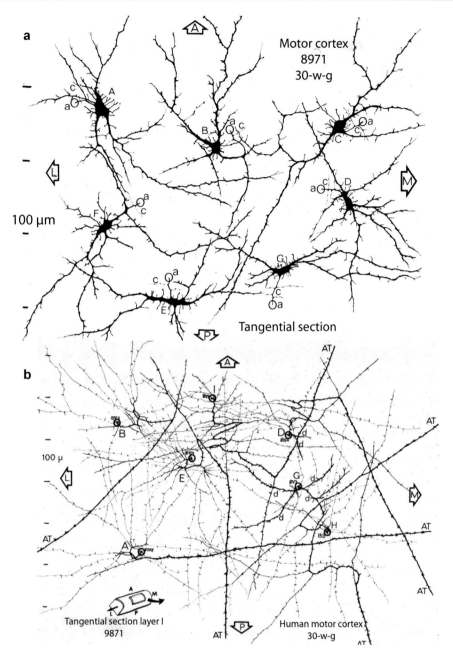

Fig. 5.10 Montage of camera lucida drawings, from rapid Golgi preparations, of the motor cortex of a 30-week-gestation human fetus, showing tangential views (**A**) of several C-R cells variable multipolar dendritic morphology and also tangential views (**B**) of their axonic collaterals and axons terminals multidirectional distribution. These neurons dendritic and axonic processes spatial orientation can be determined by the four arrows signaling the cortex anterior, posterior, medial, and lateral directions. The other features in the photos are self-explanatory

A detail camera lucida montage, from tangentially cut Golgi preparations of a 30-week-old human fetus motor cortex, illustrates the morphological and functional features of a single C-R neuron (Fig. 5.12). This camera lucida montage has been reconstructed by the superposition of three (upper, middle, and lower) separate drawings made on transparent glass of a single C-R cell. The upper one illustrates the neuron body and the length and multipolar distribution of its various dendrites, the middle one illustrates the neuron descending

5.3 First Lamina Secondary Components

The first lamina secondary components, incorporated later in development, include: a special type of astrocyte and a variety of new axonic terminals (Marín-Padilla and T. Marín-Padilla 1982; Marín-Padilla 1990). The new axonic terminals that reach the developing first lamina include: the axonic bouquets of Martinotti cells from the different cortical strata and the axonic terminals from afferent thalamic, callosal (interhemispheric), and cortico-cortical fibers (Figs. 5.2B, C, 5.3A, and 5.7A, B).

Early in neocortical development, the SP zone pyramidal-like and the primordial Martinotti neurons dendritic and axonic terminals, respectively, reach the first lamina forming bouquets that mimics that of pyramidal neurons. By the 16th week of gestation, the dendritic and axonic terminals of these early primordial neurons began to retract and to lose their first lamina contact (Chapter 3). New Martinotti neurons became progressively incorporated into the various strata of the developing cortex paralleling the ascending maturation of its pyramidal neurons (see Chapter, 3). The new Martinotti cells axonic terminals mimic those of pyramidal neurons' terminal dendritic bouquets of similar cortical depth or strata, which possibly represent their functional targets. By the time of birth, the axonic terminals of Martinotti cells are quite prominent, including those of large (deep), medium (intermediate), and small (superficial) cells (Figs. 5.1D and 5.2B, C).

A variety of small neurons with local functional fields are also progressively incorporated into the first lamina (Marín-Padilla and T, Marín-Padilla, 1982). They start to be recognized as early as the 24th week of gestation. Most of them are small stellate neurons with variable dendritic morphologies and local axonic distribution. Others have a concentrated axonic distribution, while still others resemble glial cells with short, undifferentiated, and tortuous branches. They are found scattered through all levels of the lamina and all have restricted functional territories. Both Retzius and Cajal provided excellent Golgi descriptions of these small neurons (Retzius 1894; Cajal 1911).

Another first lamina component is the ephemeral subpial granular layer (SGL) of Ranke (Ranke 1909).

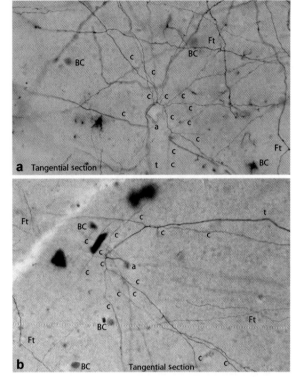

Fig. 5.11 Montage of photomicrographs, from tangential rapid Golgi preparation, of a 30- week-gestation human fetus illustrating the multidirectional expansion of two C-R cell axons (**A** and **B**), including the distribution of their axonic collaterals (c) and axon terminal (t). A few crisscrossing tangential fibers (Ft) are also stained (see also Fig. 5.10B)

axon with its 26 radiating collaterals, and the lower one the neuron single axonic terminal directional growth. The drawing also illustrates the four basic directions of the neocortex. The drawing illustrates the neuron multipolar, tangential, and horizontal dendritic morphology, the multidirectional and circular expansion of its 26 horizontal axonic collaterals, and the final orientation of its horizontal axonic terminal. The number assigned to each axonic collateral indicates the order of its emergence from the neuron descending axon through the lamina middle stratum (Fig. 5.12). During development, the functional territory of contiguous, although greatly separated, C-R cells will eventually overlap with each other covering the entire surface of the neocortex and contacting all terminal dendrites of pyramidal neurons

5.3 First Lamina Secondary Components

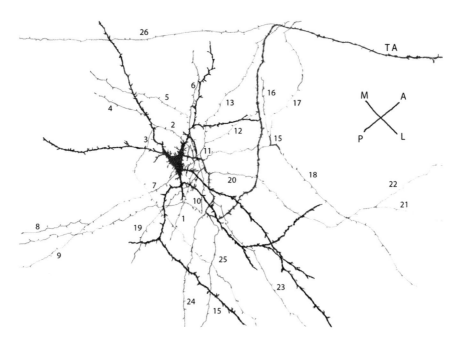

Fig. 5.12 Camera lucida montage, from tangential rapid Golgi preparations of the motor cortex of a 30-week-old human fetus, illustrating a single Cajal–Retzius neuron basic features and spatial orientation. The montage has been constructed by the superposition of three separate drawings made on transparent glass reproducing the neuron entire functional territory. The first one reproduces the first lamina upper level with the neuron body and the size and multipolar distribution of its dendrites. The second one reproduces the first lamina middle level showing the neuron descending axon with its 26 radiating axonic collaterals. The number assigned to each collateral reflects the order of its emergence from the descending axon. These axonic collaterals radiate in all direction covering a roughly circular functional territory, which will expand progressively accompanying the incorporation and centrifugal expansion of terminal dendrites of pyramidal cells, during the cortex developmental growth. The third one reproduces the first lamina lower level with the neuron axonic fiber terminal (TA) and its directional growth. The overall orientation of the neuron and the directional growth of its axon are determined by the four basic directions of the cortex(inset)

This transitional layer, composed of small undifferentiated cells, start to form throughout the first lamina lower stratum, around the 15th week, is fully developed by the 25th week, start to disappeared by the 26th week, and by the 35th week of gestation is no longer recognizable. In my opinion, SPL of Ranke undifferentiated cells are the precursors of both first lamina special astrocytes and gray matter protoplasmic astrocytes (Marin-Padilla 1995). During late prenatal and postnatal developments, the first lamina special astrocytes play an increasing role in the neocortex EGLM maintenance and repair as they progressively replace the radial glia fibers. The first lamina special astrocytes began to be recognized between the 15th and the 20th week of gestation. The origin and prenatal development of both first lamina special astrocytes and gray matter protoplasmic astrocytes are closely interrelated.

Their developmental histories are explored in detail in Chapter 8.

During late neocortical development, axonic terminals from thalamic, interhemispheric, and cortico-cortical fibers also reach and are progressively incorporated into first lamina (Figs. 5.2B and 5.3). Their functional targets are also the terminal dendrites of the pyramidal neurons. However, their distribution within the first lamina is local and, probably, specific for each different cortical region. The axonic terminals from thalamic, interhemispheric, and cortico-cortical fibers that reach the developing motor, sensory motor, visual, temporal, frontal, and parietal regions (as well as other regions), probably have different sources. These axonic terminals destination, local expansion, and possibly restricted functional influence within first lamina, are probably specific for the different cortical regions. Their local distribution and hence restricted functional territories contrast sharply with the universal and common functional field established by the C-R cells axonic processes on terminal dendrites of the pyramidal neurons throughout the neocortex. Probably, the functional roles of these late-arriving corticipetal fibers represent postnatal events in the maturation of the cortex.

References

Cajal SR (1891) Sur la structure de l'ecorce cerebrale de quelques mammiferes. La Cellule 7:125–176

Cajal SR (1897) Las células de cilindro-eje corto en la capa molecular del cerebro. Revista Trimestrar Micrográfica 2: 105–127

Cajal SR (1911) *Histologie du Systéme Nerveux de l'Homme et des Vertebrés* (reprinted: Consejo Superior Investigaciones Científicas, Madrid, 1952). Malone, Paris

Kölliker, A von, (1896). *Handbuch der Gewebelehere des Menschen.* Vol. 2. Nervensystem des Menschen und der Thiere. Leipzig, Engelman.

Marín-Padilla M (1971) Early prenatal ontogenesis of the cerebral cortex (neocortex) of the cat (Felix domestica). A Golgi study. Part I. The primordial neocortical organization. Zeitschrift für Anatomie und Entwicklungeschichte 136:117–145

Marín-Padilla M (1972) Prenatal ontogenesis of the principal neurons of the neocortex of the cat (Felis domestica). A Golgi study. Part II. Developmental differences and their significances. Z Anat Entwickl-Gesch 136:125–142

Marín-Padilla M (1978) Dual origin of the mammalian neocortex and evolution of the cortical plate. Anatomy and Embryology 152:109–126

Marín-Padilla M, Marín-Padilla T (1982) Origin, prenatal development and structural organization of layer I of the human cerebral (motor) cortex A Golgi Study. Anatomy and Embryology 164:161–206

Marín-Padilla M (1983) Structural organization of the human cerebral cortex prior to the appearance of the cortical plate. Anatomy and Embryology 168:21–40

Marín-Padilla M (1990) Three-dimensional structural organization of layer I of the human cerebral cortex: A Golgi study. Journal of Comparative Neurology 229:137–146

Marín-Padilla M (1992) Ontogenesis of the piramidal cell of the mammalian neocortex and developmental cytoarchitectonics: A unifying theory. Journal of Comparative Neurology 321: 223–240

Marín-Padilla M (1995) Prenatal development of fibrous (white matter), protoplasmic (gray matter) and layer I astrocytes in the human cerebral cortex: A Golgi study. Journal of Comparative Neurology 358:1–19

Marín-Padilla M (1996) Developmental neuropathology and impact of perinatal brain damageI, Hemorrhagic lesions of the neocortex, Journal of Neuropathology and Experimental Neurology 55:746–762

Marín-Padilla M (1997) Developmental neuropathology and impact of perinatal brain damageII. White matter lesions of the neocortex. Journal of Neuropathology and Experimental Neurology 56:219–235

Marín-Padilla M (1998) The Cajal–Retzius cell and the development of the neocortex. Trends in Neurosciences (TINS) 21:64–71

Marín-Padilla M (1999) Developmental neuropathology and impact of perinatal brain damage. III, Gray matter lesions of the neocortex. Journal of Neuropathology and Experimental Neurology 58:407–429

Marín-Padilla M, Parisi JE, Amsrtrong DL, Sargent SK, Kaplan JA (2002) Shaken infant syndrome: Developmental neuropathology, progressive cortical dysplasia and epilepsy. Acta Neuropatholgica 103:321–332

Martin R, Gutierrez A, Pefiafiel A, Marin-Padilla M (1999) Persistence of Cajal–Retzius cells in the adult human cerebral cortex. An immunohistochemical study. Histol Histopathol 14:487–490

Ranke O (1909) Beitrage zur Kenntnis der normalen und pathologischen Hirnrinden Baldung. Beitr Pathologie und Anatomie 45:51–85

Retzius G (1893) Die Cajal'schen Zellen der Groshirnrinde baim Menschen und bei Säugentieren. Biology Unters 5:1–9

Retzius G (1894) Weitere Beiträge zur Kenntniss der Cajal'schen Zellen der Grosshirnrinde des Menschen. Biology Unters 6: 9–34

Rocker LB, Riggs HE (1969) *Myelinization of the brain in the newborn.* Lippincott, Philadelphia

Yakovlev PI, Lecours AR (1967) The myelogenic cycles of regional maturation of the brain. In: Minkowski A (ed) *Regional Development of the Brain in Early Life.* Blackwell Publishers, Oxford, pp 3–70

Human Motor Cortex Excitatory–Inhibitory Neuronal Systems: Development and Cytoarchitecture

An essential feature of the mammalian cortex is the sequential establishment of excitatory–inhibitory neuronal assemblages between pyramidal and local-circuit inhibitory neurons. The recognition, location, and distribution of these excitatory–inhibitory neuronal assemblages are fundamental objectives in the study of the nervous system. They are recognized in all strata of the motor cortex. Most local-circuit interneurons recognized in the motor cortex are inhibitory in nature. Because, pyramidal cells represent roughly the 70% and the local-circuit interneurons the 30% of the motor cortex gray matter neurons, each inhibitory neuron establishes synaptic contacts with numerous pyramidal neurons. The excitatory pyramidal neurons are morphologically stable and functionally anchored to the first lamina. On the other hand, the inhibitory neurons are free, without first lamina attachment, capable of modifying their dendritic and axonic profiles as well as their spatial distribution in response to learned (acquired) new motor activities. Excitatory–inhibitory functional systems are essential components of the cerebral cortex normal function, and possibly of abnormal and or altered functional activity (Lewis et al. 2005). The information on inhibitory interneurons already available in the literature is considerable. In this chapter describes the developmental histories of four basic excitatory–inhibitory systems (Martinotti, basket, double-tufted, and chandelier) of the human motor cortex, using the rapid Golgi procedure. While all pyramidal neurons of the cortex are generated in the ependymal epithelium, the origin of inhibitory interneurons seems to be extracortical. Inhibitory neurons enter into the developing cerebral cortex advancing horizontally, following a proximal (ventral) to distal (dorsal) gradient and become sequentially incorporated into it, paralleling the ascending maturation of the various pyramidal cell strata (Chapter 3). In the developing human motor cortex, horizontally traversing neuronal precursors are recognized as early as the 15th week of gestation (Chapter 3). The morphology of horizontally traversing neurons is undefined and cannot be identified as distinct types. The recognition of the inhibitory neurons distinctive morphologies occurs later in development and coincides with ascending maturation of the pyramidal neurons of the various strata (Chapter 3).

Therefore, following their incorporation into the developing motor cortex, these neurons go through two stages: an early undifferentiated one as they become incorporated into a specific pyramidal cell strata and a later one when they acquire distinctive morphological and functional features. The motor cortex inhibitory interneurons are characterized by distinctive dendritic profiles, by specific axonic terminals on pyramidal neurons and by wanting first lamina dendritic attachment. Their recognition parallels the ascending functional maturation of the various pyramidal cell strata with which they establish functional contacts. They are first recognized through the lower and older pyramidal cell strata and subsequently throughout upper and younger ones. While they are recognized in lower pyramidal cell strata, they remain undifferentiated and unrecognizable in upper and younger ones. By the time of birth, various types of inhibitory local circuit interneurons are recognized throughout all-pyramidal cell strata of the motor cortex (Marín-Padilla 1969, 1990).

From the 8th to the 16th week of gestation, only pyramidal neurons anchored to first lamina by their terminal dendrites are recognized in the motor cortex. From the 16th to the 22nd week, small local-circuit neurons, without distinctive dendritic and/or axonic features, start to be recognized among the pyramidal neurons of the lower strata. At these ages, no inhibitory neurons are recognized in the upper, younger, and still undifferentiated pyramidal cell strata. Subsequently and up to birth time, four types of inhibitory neurons (Martinotti, basket, double-bouquet, and chandelier) are recognized in the human motor cortex following an ascending maturation that parallels that of pyramidal neurons.

Four distinct excitatory–inhibitory systems have been, so far, recognized in the developing human motor cortex. They include the pyramidal-basket, the pyramidal-Martinotti, the pyramidal-double-bouquet, and the pyramidal-chandelier systems, respectively. Each one is characterized by distinctive dendritic profiles, specific axonic terminals on pyramidal neurons and by a 3-D (spatial) distribution within the motor cortex. Their functional targets on pyramidal neurons include: the terminal dendritic bouquets for Martinotti cells, the soma for basket cells, the apical dendritic shaft for double-tufted cells, and the axon-hillock for chandelier cells. The developmental histories, morphological, and functional features of these four excitatory–inhibitory neuronal systems are described below. Since most inhibitory neurons appear to be spatially oriented, the rapid Golgi preparations need to be cut parallel, perpendicular, and tangential to the gyrus long axis.

It should be emphasized that there are additional short-circuit interneurons throughout the cerebral cortex as well as additional excitatory–inhibitory systems, which are not yet fully identified.

6.1 The Pyramidal-Basket System

Undoubtedly, the best-known excitatory–inhibitory system is that formed between pyramidal neurons and their corresponding inhibitory basket cells. This unique neuronal association was originally described in the mammalian cerebellum, hippocampus, and cerebral cortex (Cajal 1911; Andersen, Eccles and Løyning 1963a; Andersen et al. 1963b; Eccles 1964; Szentágothai 1965, 1975, 1978; Marín-Padilla 1969, 1990). Cajal described the formation of pericellular nests (baskets) around the cell body of cerebellar Purkinje cells, of hippocampus pyramidal cells and, eventually, on pyramidal neurons of the human visual and motor cortex (Cajal 1893, 1899). In the cerebellum and hippocampus, Cajal demonstrated the association of the pericellular baskets with a distinct type of local-circuit stellate interneuron; but did not describe the original interneuron that formed the baskets around the pyramidal neurons in the cerebral cortex (Cajal 1911). However, he suggested that large stellate neurons with horizontal axonic collaterals contiguous to pyramidal neurons could by the source of the baskets (Cajal 1911). In 1965, Szentágothai, using chronically isolated cortical slabs, with intact microvasculature, demonstrated the inhibitory function of the pericellular nest around pyramidal neurons and the local distribution of the inhibitory neurons forming them (Szentágothai 1965, 1975, 1978). Marín-Padilla, using rapid Golgi preparations, described a distinct type of stellate interneuron that form pericellular basket around the body of pyramidal neurons of the human motor and visual cortex, which since them, they are recognized as basket cells (Marín-Padilla 1969, 1970, 1972; Jones and Hendry 1984). The axonic terminals of pericellular baskets were retrograded up to these stellate interneurons and these neurons' axonic terminals were followed up to the baskets. The pyramidal-basket assemblage is the best-studied excitatory–inhibitory system of the cerebral cortex, including that of humans (Eccles 1964; Marín-Padilla 1969, Jones 1984; Marín-Padilla Jones and Hendry 1984; 1970, 1972, 1974, 1990).

The Basket Cell. The association of large stellate interneurons with large pyramidal neurons is a distinctive feature of the human motor cortex (Fig. 6.1). Basket cells are local-circuit stellate inhibitory interneurons characterized by ascending, descending, and horizontal dendrites and by an ascending and/or descending axon that branch, at various levels, into multiple long horizontal collaterals. From the long horizontal collaterals emerge numerous shorter branches that participate in the formation of the pericellular nests (baskets) around the pyramidal cells bodies (Figs. 6.1–6.3). The presence of distinct stellate basket cells among pyramidal neurons is recognized in the human motor cortex from the 30th week of gestation up to birth time. Their arrival into the developing motor cortex precedes this gestational age. At birth, these inhibitory interneurons have spiny dendrites. The terminals arising from the neuron long horizontal axonic collaterals participate in the formation of

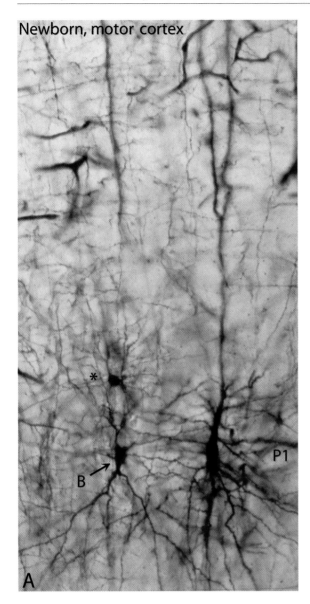

Fig. 6.1 Photomicrograph from a rapid Golgi preparation of the newborn infant motor cortex showing, comparatively, the size and dendritic profiles of a large pyramidal neuron of stratum P1 with apical dendrite reaching first lamina, a giant stellate basket cell (B) of the same stratum and a smaller (*) undetermined neuron. There are also numerous vertical and horizontal fiber terminals

cell strata. Their size and location varies paralleling that of its associated pyramidal neurons (Marín-Padilla 1972). In the newborn human motor cortex, basket cells of various sizes are recognized, including: giant (P1 stratum), large (P2), medium (P3–P4), and small (P5–P6) basket cells, respectively (Marín-Padilla 1970, 1990). Regardless of their different sizes and/or location they all share similar dendritic and axonic profiles, reflecting that of its associated pyramidal neurons (Figs. 6.1–6.3).

The functional territory of a single basket cell is quite large and extends through adjacent pyramidal cell strata (Fig. 6.2 and 6.3). The basket cell size, its functional territory, and the number of pericellular baskets formed decrease from lower to upper pyramidal cell strata (Figs. 6.2 and 6.3). In the newborn motor cortex, basket cells throughout the upper pyramidal cell strata (P6–P5) are still immature and their contributing baskets are only partially constructed (Figs. 6.2a, b). Basket cells throughout lower pyramidal cell strata (P1–P2) are larger and their contributing baskets are larger and completely formed (Fig. 6.3b, c). The basket cells and the size and complexity of their baskets continue to increase during postnatal developments (Fig. 6.3b, c).

It is also important to point out, that a basket cell can only be visualized in its entirety (dendritic and axonic profiles and terminal baskets) in rapid Golgi preparations cut perpendicular to the pial surface and to the long axis of the gyrus. Basket cells are flat stellate neurons spatially oriented perpendicular to the gyrus long axis and, consequently, their morphological appearance might change depending of the angle of view. In rapid Golgi preparations cut perpendicular to the gyrus, it is relatively easy to locate these stellate interneurons among the pyramidal neurons (Fig. 6.1). In these preparations once the basket cell soma is localized, by moving the microscope micrometer – up and down – (5 μm each time), it is possible to follow-up its dendritic and axonic arbors in their entirety and determine their number, length, distribution, as well as the number of terminal baskets formed by them. It is also possible to determine the overall thickness of the selected basket cell. By calculating the number of micrometer moves – up and down – it is possible to estimate the basket cell thickness. The overall thickness of the large basket cells of pyramidal cell strata P1 and P2 ranges between 45 and 55 μm. The basket cell thickness and the size of their

multiple pericellular baskets (Figs. 6.2 and 6.3). A single basket cell participates in the formation of pericellular baskets around pyramidal neurons of its strata and of adjacent ones. Also, each pericellular basket is formed by the contributions of axonic terminals from several basket cells from its adjacent strata (Figs. 6.2 and 6.3). Basket cells are recognized in all pyramidal

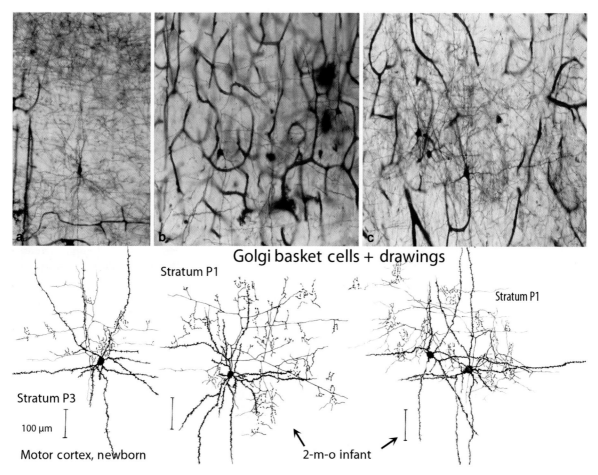

Fig. 6.2 Montage of photomicrographs and corresponding camera lucida drawings, from rapid Golgi preparations, of newborn infant's motor cortex illustrating the size, location, stellate morphology, and dendritic and axonic profiles of inhibitory basket interneurons. Basket cells from the following pyramidal cell strata are illustrated: from pyramidal strata P6 (**a**), P5 (**b**), and P4 (**c**) strata. The axon terminals of all basket cells participate in the formation of perisomatic nests (baskets); some of them are still incompletely formed

functional territories throughout the upper cortical strata are proportionally smaller.

The axonic distribution of a large basket cell could cover a rectangular functional territory that is also flat and perpendicular to the long axis of the gyrus (Figs. 6.2 and 6.3). The size of the functional territory of a large basket cell, from P1 to P2 pyramidal cell strata, measures roughly 50 μm in width, 500 μm in height, and 1,000 μm in length (Fig. 6.4b). The sizes of the functional territories of baskets cells of upper pyramidal cell strata (P3–P5) are proportionally smaller. To corroborate the basket cells spatial orientation, a computer reconstruction of a large basket cell was carried out (Fig. 6.4a). Separate and sequential camera lucida drawings, of a selected large basket cell from a 2-month-old (m-o) infant motor cortex, were made at 5 μm intervals until the entire neuron was drawn (Marín-Padilla and Stibitz 1974). Starting with the neuron body, each separate drawing records only those dendritic and axonic processes that appear in sharp focus (Fig. 6.4a). The series of incomplete drawings were digitalized individually and, with a computer program, they were reconnected to obtain a digitalized 3-D model of the neuron (Fig. 6.4a). It was then possible to rotate the digitalized basket cell and view it from different angles, ranging from 0° to 80° (Fig. 6.4a), thus confirming that basket cells have flat rectangular dendritic and axonic functional territories oriented perpendicular to the pial surface and to the long axis of the precentral gyrus (Fig. 6.4b).

6.1 The Pyramidal-Basket System

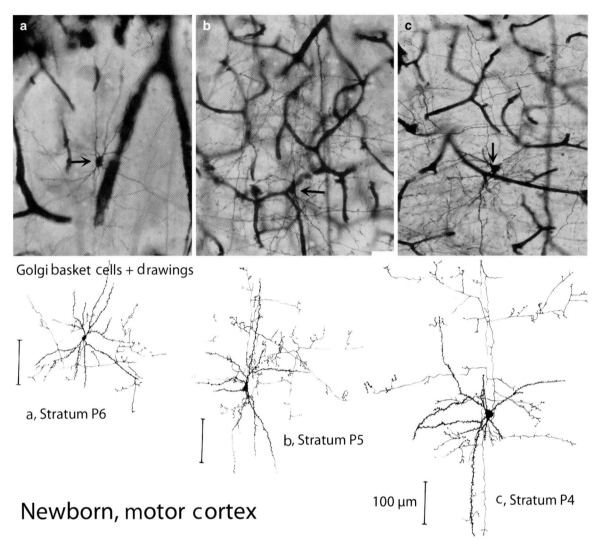

Fig. 6.3 Montage of photomicrographs and corresponding camera lucida drawings, from rapid Golgi preparations, of the motor cortex of newborns and a 2-month-old infants, illustrating the size, location, stellate morphology, and dendritic and axonic profiles of basket cells from various pyramidal cells strata, including P3 (**a**) and P1 (**b, c**) strata. The axon terminals from all basket cells participate in the formation of perisomatic baskets around pyramidal cell bodies. The number and complexity of the baskets are greater than those of the upper pyramidal cell strata (compare with those of Fig. 6.2)

Consequently, a basket cell viewed in profile (at an 80° angle) and/or in sections cut parallel to the gyrus axis assumes the morphologic features of a bipolar interneuron (Fig. 6.4a). In this context, it should be pointed out that Cajal, in his book, describes a type of double-bouquet (bipolar) neuron associated with the formation of pericellular baskets (Cajal 1911). Since the orientation of Cajal's Golgi preparations was not documented, the double-bouquet neuron he described could have actually represented a basket cell viewed in profile.

A single basket cell forms pericellular baskets around the bodies of all pyramidal neurons within its rectangular functional territory. The functional territory of any basket cell overlaps with contiguous ones, above and below it, with a similar spatial orientation (Fig. 6.4b). From superficial P6 to deep P1 pyramidal cell strata, the motor cortex is subdivided into a series of overlapping rectangular functional territories perpendicular to both pial surface and the long axis of the gyrus, which are interconnected by their basket cells (Fig. 6.4b). The motor cortex may be viewed as a series of perpendicular

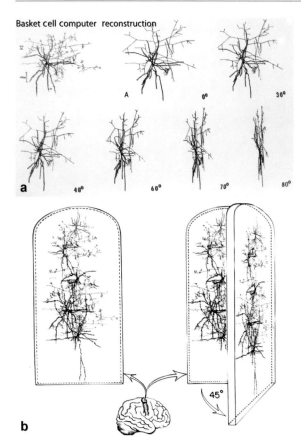

Fig. 6.4 Schematic representations of (**a**) the rotation of a computer reconstructed basket cell of P1 pyramidal cell stratum of the motor cortex of a 2-month-old infant and of (**b**) the vertical and rectangular pyramidal-basket functional assemblages established in all pyramidal cell strata of the motor cortex, which extend throughout the entire gyrus like the pages of a book. The progressive rotation of the basket cell (A) changes its stellate morphology into a bipolar one. Basket cells are flat and spatially oriented neurons and their morphologic appearance varies depending on the angle of view (see also Figs. 6.2 and 6.3).

and spatially oriented rectangular functional territories composed of excitatory–inhibitory pyramidal-basket systems extending throughout the entire precentral gyrus, like the pages of a book. The size and orientation of these pyramidal-basket functional territories are not unlike the narrow and vertical functional fields established by some specific corticipetal fibers systems (Blakemore and Tobin 1972; Hubel and Wiesel 1977; Jones 1984; Jones and Hendry 1984).

The fact that basket cells are spatially oriented and that their morphologic appearance may changes as the angle of observation changes is an important observation with significant implications. Specific spatial orientations may also be applicable to other excitatory–inhibitory systems. Possibly, all local-circuit interneurons of the mammalian cortex have spatially oriented functional territories. Therefore, the need to establish the spatial orientation of any brain section is mandatory as the appearance and distribution of some local-circuit interneurons could change depending on the angle of view. On the other hand, all pyramidal neurons, throughout the cerebral cortex, lack specific spatial orientation and their morphologic appearance remains unchanged from any angle of view. However, all pyramidal neurons throughout the cerebral cortex are interconnected by spatially oriented functional systems, represented by the inhibitory neurons of its various strata. The bland and uniform appearance of the pyramidal neurons of the cerebral cortex, from any angle of view, gain considerable functional relevance by its association with the spatially oriented inhibitory interneurons.

The C-R cells that target the pyramidal terminal dendrites throughout the neocortex represent another important and spatially oriented functional system of the cerebral cortex (Chapter 5).

The mammalian neocortex can be thought of as a complex organization composed of superimposed strata of fixed (unchangeable) excitatory pyramidal neurons bounded and functionally interconnected by spatially oriented inhibitory systems capable of establishing functional contacts only with neurons within their functional territory, such that, a single pyramidal cell column may receive functional contacts from differently oriented inhibitory interneurons. In this context, the motor cortex may be thought of as a forest (paraphrasing Cajal's idea) with innumerable immovable tree trunks (pyramidal cells apical dendrites) and by a series of moving shadows that make contacts with different tree trunks. As the shadows move, different tree trunks become contacted. While the immovable tree trunks represent the neocortex immovable projective pyramidal neurons, the moving shadows represent the spatially oriented local-circuit inhibitory neurons. During the subsequent development and functional maturation of the neocortex, the numbers of different spatially oriented inhibitory neuronal systems contacting pyramidal neurons and their extent could be extraordinary and unimaginable, thus multiplying many folds the functional possibilities of the same neurons. The need for further explorations into the morphology and function of these inhibitory neuronal are important goals in the study of the human brain.

6.1 The Pyramidal-Basket System

The Pericellular Nest or Basket. The presence of stellate interneurons among the large pyramidal neurons of P1 stratum is first recognized around the 25th week of gestation. However, at this stage, the pericellular baskets are not yet recognizable as distinct structures and these interneurons have not yet developed distinct dendritic profiles. During the neocortex subsequent ascending functional stratification, stellate neurons (no yet fully developed) forming terminals pericellular baskets are progressively recognized among the ascending maturation of the various pyramidal cells strata. Fully developed pericellular baskets have been described in the human motor and visual cortex of newborn infants, using the rapid Golgi procedure (Marín-Padilla 1969, 1970, 1972, 1974, 1990).

By the time of birth, pericellular baskets are recognized throughout all pyramidal cell strata (Figs. 6.2 and 6.3). A pericellular basket is a prominent, triangular shaped, and complex structure composed of numerous axonic terminals that establish axo-somatic contacts with the pyramidal cell body (Fig. 6.5a, b). The triangular (pyramidal) shapes of baskets mimic that of pyramidal cell bodies (Figs. 6.6–6.8). It is important to point out that, in rapid Golgi preparations, a pericellular basket can only be visualized if the pyramidal cell body is not stained (Figs. 6.5a, b and 6.6a, b). In rapid Golgi preparations that the pyramidal cell body is stained, the pericellular baskets will be unrecognizable (Fig. 6.1; and Figs. 4.1, 4.9 and 9.3). Contrarily, if the pyramidal cell body is not stained the entire pericellular basket can be visualized (Figs. 6.6–6.9). This rapid Golgi procedure welcome idiosyncrasy has permitted the visualization of the entire pericellular basket and has facilitated its 3-D reconstruction. This welcome rapid Golgi behavior remains inexplicable and cannot be preplanned, which further supports the idea of making as many Golgi preparations as is possible. Perhaps one of those preparations could offer an unexpected but extraordinary view of the nervous tissue (Chapter 12).

At low magnification, the pericellular baskets throughout the lower pyramidal cell strata (P1 and P2) are quite prominent and contact sharply with the surrounding tissue (Figs. 6.5a, b). The concentration of pericellular baskets throughout the upper pyramidal cell strata is less conspicuous (Figs. 6.2 and 6.3). The staining of the pyramidal cell body will obscure the whole pericellular axonic nest visualizing only its lateral elements. Contrarily, if the neurons body is unstained, the pericellular basket anterior, laterals and posterior walls

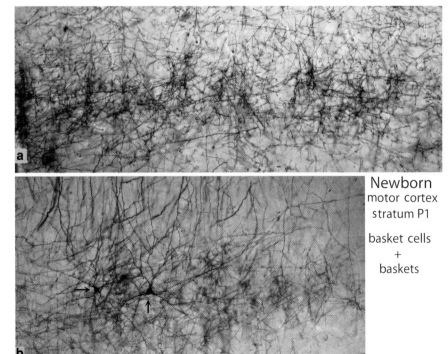

Fig. 6.5 Montage of photomicrographs, from rapid Golgi preparation of newborn infant's motor cortex, showing (**a**) the number and structural complexity of pericellular baskets from around the bodies of unstained pyramidal neurons of stratum P1 and of a few basket cells (**b**) associated with them. In Golgi preparations, pericellular baskets can only be visualized if the pyramidal neurons are unstained. Numerous axonic terminals from basket cells concentrate around the unstained pyramidal cell somata forming roughly triangular-shaped complex pericellular nests or baskets that mimic the neuron's body size and shape. Both illustrations also show numerous fine horizontal fiber terminals representing basket cells horizontal axonic fibers

Newborn motor cortex stratum P1 basket cells + baskets

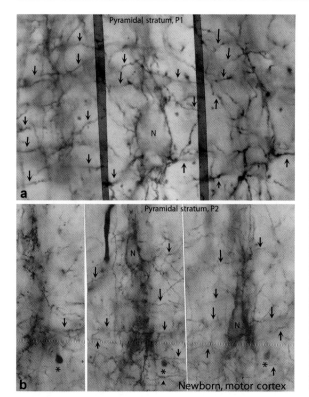

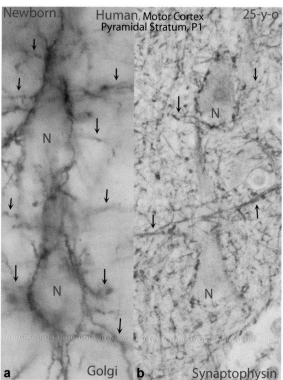

Fig. 6.6 Montage of photomicrographs, from rapid Golgi preparations of the motor cortex of newborn infants, showing, at a higher magnification, three levels (superior, middle, and inferior) of pericellular baskets formed around the unstained body of P1 (**a**) and of P2 (**b**) pyramidal neurons. The illustrations demonstrate the baskets 3-D structural organization, its fiber terminals complexity, and its apparently empty center occupied by the pyramidal neuron unstained (N) body. Arrows identify the large number of horizontal axonic terminals that participate in the basket formation. Asterisks (*) identify a blood capillary coursing near the basket (**b**). The illustrations offer three views of the same basket, which covers a vertical and rectangular area, roughly 45–55 µm thick. The illustrations also demonstrate that good rapid Golgi preparations permits very high resolution without losing the fine structural details of the nervous tissue, in clarity as well as in authenticity

Fig. 6.7 Montage of photomicrographs comparing, at a very high magnification, the structural complexity and remarkable similarities of pericellular baskets from the motor cortex of (**a**) a newborn infant (rapid Golgi preparations) and (**b**) from a 25-year-old man (synaptophysin stain preparations). The arrows indicate some of the horizontal fiber terminals that participate in the basket formation. The baskets apparently empty center is occupied by the unstained pyramidal cells (N) bodies (Golgi preparations), which are partially stained in the synaptophysin stain

are clearly recognized (Figs. 6.6a, b, 6.7a, b, and 6.8a–c). In good rapid Golgi preparations, it is possible to demonstrate the three-dimensional complex organization of the basket and its apparently empty center occupied by the unstained pyramidal cell body (Figs. 6.6a, b, 6.7a, and 6.8a–c). Separate photomicrographs taken above, center, and below a pericellular basket demonstrate its structural complexity, the numerous –terminal, and en-passant-axo-somatic contacts and its apparently empty center (Figs. 6.6–6.8). Each basket is composed of axonic terminals from several adjacent basket cells and any basket cell provides axo-somatic contacts to several pyramidal cell bodies (Figs. 6.6a, b and 6.7a, arrows). The structural organization of a pericellular basket can also be demonstrated, in the adult cerebral cortex, using synaptophysin stain preparations (Fig. 6.7b). These preparations demonstrate the basket numerous axo-somatic lateral contacts around the pyramidal cell body but fail to demonstrate its overall 3-D organization (Fig. 6.7b).

Montages of camera lucida drawings can be made by superimposing three separate ones taken from the basket upper, middle, and lower levels (Fig. 6.8a–c). These montages demonstrate the basket extraordinary structural complexity, the number of axonic terminals reaching it, the number axo-somatic (terminals and in-passant) contacts, as well as their apparently empty center (Fig. 6.8a–c). Electron microscopic and immunocytochemical studies have confirmed the basket's

6.2 The Pyramidal-Martinotti System

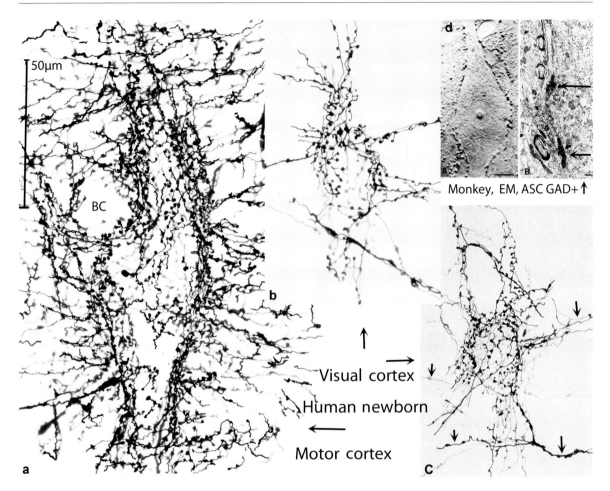

Fig. 6.8 Montage (**a, b, c**) of camera lucida drawings, showing the basket 3-D structural organization, from rapid Golgi preparations of the motor (**a**) and the visual (**b, c**) cortex of newborn infants. Superimposing three separate drawings of each basket anterior, middle, and posterior levels makes the illustrations. The reconstructions also show the numerous, terminal and en-passant, synaptic axo-somatic contacts established on each basket. (**d**) Illustrate an electron microscopic and immunohistochemical (ASC and GAD+) views of baskets from the monkey cerebral cortex showing the numerous axo-somatic synaptic contacts of the pyramidal neuron body as well as their inhibitory nature. (From Jones and Hendry 1984)

numerous axo-somatic contacts as well as their inhibitory nature (Jones and Hendry 1984). These studies, however, failed to convey the basket's 3-D structural organization (Fig. 6.8d), a fact that further encourages the need to use the rapid Golgi procedure to study of the brain's cytoarchitectural organization.

6.2 The Pyramidal-Martinotti System

Martinotti cells with ascending axons that reach and branch within the first lamina are among the earlier recognized local-circuit interneuron in the mammalian developing neocortex (Chapters 2 and 3). These early Martinotti cells are essential components of the neocortex subplate (SP) zone. Together with the SP pyramidal-like neurons and first lamina Cajal–Retzius (C-R) cells, they represent the elements of the primordial cortical organization, prior to the appearance of the mammalian pyramidal cell plate (PCP) (Chapters 2, 3, and 5). Their ascending axon fans into a terminal bouquet that mimics the dendritic bouquets of SP pyramidal-like neurons. These early Martinotti cells are local-circuit interneurons, which interconnect structurally and functionally the SP zone pyramidal-like neurons and the first lamina C-R neurons.

These early SP Martinotti cells are recognized up to around the 15th week of gestation (Chapter 3). Afterward, their axonic terminals start to lose their first lamina contacts and to regress. During subsequent development, these early SP interneurons (together with pyramidal-like neurons) are progressively displaced downward and transformed into deep interstitial neurons. Subsequently, new Martinotti inhibitory interneurons are progressively incorporated into the cortex PCP paralleling the ascending maturation of its pyramidal cell strata. Martinotti cells are bipolar interneurons characterized by long ascending and descending dendrites with spine-like excrescences and by a long ascending axon that crosses the maturing PCP and reaches the first lamina (see Fig. 3.14). Because of their length, the deep (P1 stratum) Martinotti cells with long axon are difficult to see in a single Golgi preparation, although their location, dendritic profiles, and the axon proximal segment can be recognized at various pyramidal cells strata. Similarly their distinct terminal axonic bouquets are also easily recognizable within the first lamina (Fig. 6.9a–c).

The most distinguishing feature of these local-circuit neurons is their ascending axon that reaches and fan into the first lamina forming a terminal bouquet with spine-like projections (Fig. 6.9a–c and Fig. 3.14a). These neurons terminal axonic bouquets mimic the size of the dendritic bouquets of pyramidal neurons of their own strata, which suggest both structural as well as functional

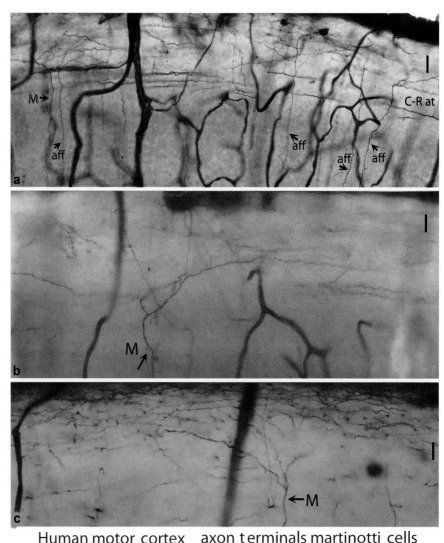

Fig. 6.9 Montage of photomicrographs, from Golgi preparations, of the motor cortex of 29-week-gestation fetuses (**a**, **b**) and of a newborn infant (**c**) showing the distinctive terminal axonic bouquets of deep Martinotti neurons (M) as they branch within first lamina (I), some afferent (aff) fiber terminals (**a**) and the C-R cells axon terminals (C-R at). The size and distribution of Martinotti neurons terminal axonic bouquets mimic the terminal dendritic bouquets of pyramidal neurons of their cortical stratum

interrelationships between them (Fig. 6.9b, c). The new Martinotti neurons are first recognized through the lower pyramidal cells strata and subsequently, during prenatal development, in upper strata. By birth time, Martinotti neurons are recognized in most pyramidal neurons strata (Marín-Padilla 1984). Their size and extent of their terminal axonic bouquets varies with their cortical location, mimicking that of corresponding pyramidal neurons of similar cortical depth. Deep-sited Martinotti neurons form larger axonic terminal bouquets than superficial ones, further suggesting structural as well as functional interrelationship between them.

Probably, Martinotti neurons terminal axonic bouquets establish inhibitory synaptic contacts with the dendritic bouquets of the corresponding pyramidal neurons. Consequently, the functional territory of a single Martinotti cell is local and limited to the terminal dendritic bouquets of pyramidal neurons from its cortical strata. This excitatory–inhibitory neuronal system is the first one established and recognized in cortical development. Additional pyramidal-Martinotti systems are progressively incorporated during the ascending maturation of its various pyramidal cells strata (Marín-Padilla 1984).

6.3 The Pyramidal-Double-Bouquet System

Double-bouquet (Cajal bipenachada) local-circuit interneurons are the most frequently recognized and described inhibitory neurons in the mammalian neocortex (Cajal 1911; Somogyi and Cowey 1984). In rapid Golgi preparations cut perpendicular to the pial surface and the long axis of the precentral gyrus, these cells appears as bipolar neuron characterized by several long and closely arranged ascending and descending dendrites (Figs. 6.10a–d and 6.11a, b). Their size varies according to their location and/or cortical strata: in the newborn motor cortex, ranging from small of P5 stratum (Figs. 6.10a), to medium of P4 stratum (Fig. 6.10b, c), to large of P3–P2 strata (Figs. 6.10d and 6.11a) to giant of P1 stratum (Fig. 6.11c). Their size corresponds to that of pyramidal neurons of similar cortical strata. The incorporation, recognition, and morphological and functional maturations follow as ascending and stratified progression that parallels that of the various pyramidal cell strata.

The neuron primary dendrites emerge from the top and bottom poles of their body and branch into several long ascending and/or descending cascading branches (Figs. 6.10 and 6.11). The dendrites are closely arranged and extend for a considerable distance on either direction. The length and distribution of the neuron ascending and descending dendrites are similar, appearing as mirror image of each other (Figs. 6.10a–d and 6.11a, b). These neurons are the most commonly recognized in the developing human motor cortex. The morphology of many still undifferentiated local-circuit interneurons is essentially bipolar.

These interneurons axonic terminal profiles are also quite characteristic. The axon emerges from the neuronal body and branches into a series of long and cascading ascending and descending terminals, which are also closely arranged (Figs. 6.10a–d and 6.11a, b). Their axonic distribution covers a vertical cylindrical functional territory. The distance between contiguous axonic terminals corresponds, roughly, with the thickness and the separation of the apical dendrites of its associate pyramidal neurons (Fig. 6.11c). The functional territories of these interneurons vary in size, and decreases from lower to upper cortical strata. Moreover, the distance between its long axonic terminals also decreases from lower to upper strata reflecting the size and thickness of the apical dendrites of their corresponding pyramidal neurons (Figs. 6.10 and 6.11). In rapid Golgi preparations (as in the case of basket cells), if the pyramidal neuron's apical dendrites are stained, the double-bouquet neurons axonic terminals will not be adequately visualized or recognized. Contrarily, if the apical dendrites of pyramidal neurons are not stained, the long cascading axonic terminals of these interneurons are clearly visible (Fig. 6.11c). Their ascending and descending axonic processes have spine-like excrescences (Fig. 6.11c, arrows). These excrescences will facilitate direct contact with the dendritic shaft, between the numerous spines. These axonic excrescences could represent the double-bouquet interneurons with specialized presynaptic units (Fig. 6.11c).

By the time of birth, double-bouquet interneurons are recognized in all pyramidal cell strata (Figs. 6.10a–d and 6.11a, b). Both, the interneuron size and that of its functional territory increase progressively from upper to lower pyramidal cell strata, paralleling that of their pyramidal neurons functional cohorts (Figs. 6.10 and 6.11). Based on rapid Golgi observations, the functional target of each double-bouquet interneuron

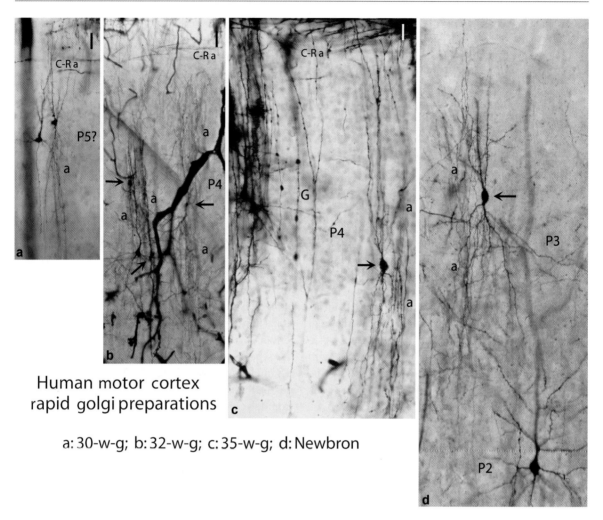

Fig. 6.10 Montage of photomicrographs, from rapid Golgi preparations of the motor cortex of 30- (**a**), 32- (**b**) and 32-week-old (**c**) fetuses and of a newborn infant (**d**), illustrating the location and morphologic features of double-tufted inhibitory interneurons (*arrows*) of pyramidal cell strata P5 (**a**), P4 (**b**), P4 (**c**), and P3 (**d**), respectively. Also illustrated are the C-R cells (C-R a) horizontal axon terminals within first lamina (**a–c**), the double-tufted neurons ascending and descending axonic terminals cascades (**a–b**), and pyramidal neurons of stratum P2 (**a**), P4 (**c**), and P2 (**d**). The size and functional territories of double-tufted interneuron increases from upper to lower pyramidal cell strata paralleling that of the pyramidal neurons of their corresponding strata

appears to be the apical dendritic shaft of several contiguous pyramidal neurons of the same strata. The double-bouquet neurons functional territory covered by its ascending and descending axonic terminals is vertical and cylindrical. Each one may include a cluster of 6–12 apical dendrites of neighboring pyramidal neurons. Moreover, in tangential rapid Golgi preparations of the human motor cortex, the presence of small (6–8) and large (10–12) apical dendrites clusters is quite common finding. The size of these tangentially cut cylindrical clusters of apical dendrites also decreases from lower to upper pyramidal cell strata and may overlap with each other, throughout the motor cortex. The diameter of the apical dendrites within these clusters also decreases from lower to upper pyramidal cell strata. It is important to point out that the morphologic appearance of double-bouquet neurons remains essentially unchanged in rapid Golgi preparation cut parallel and/or perpendicular to the precentral gyrus long axis.

Therefore, the human motor cortex pyramidal neurons are also interconnected by a series of overlapping cylindrical (columnar) functional territories established by the double-bouquet inhibitory interneurons. These

6.4 The Pyramidal-Chandelier System

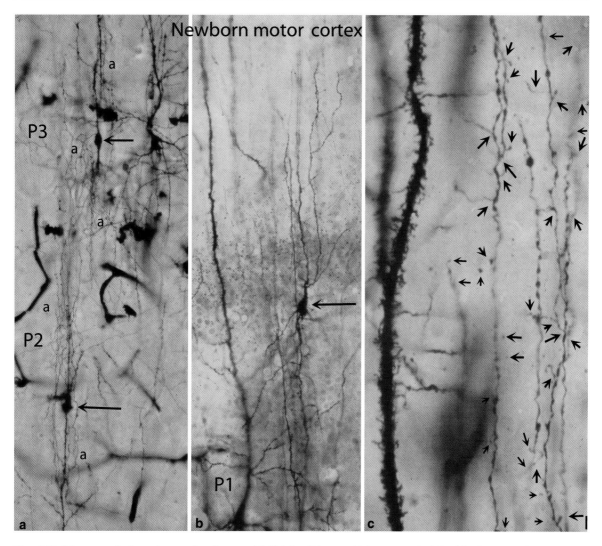

Fig. 6.11 Montage of photomicrographs, from rapid Golgi preparations of newborn infants, illustrating the size, location, and morphologic features of double-tufted inhibitory interneurons of pyramidal cell strata P3 and P2 (**a**) and P1 (**b**). Also illustrated, at a high magnification (**c**), are these interneuron's vertically and closely arranged axonic terminals with numerous spines-like short projections (*arrows*) as well as a segment of a P1 pyramidal neuron apical dendrite for comparison. The distance between these interneurons vertical axonic terminals is roughly comparable to the thickness of pyramidal neurons apical dendrites (**c**) of similar cortical strata. The double-tufted interneurons axonic terminals inter-distance decreases from lower to upper cortical strata reflecting their decreasing size as well as that of their functional cohorts, the pyramidal neurons

columnar functional territories extend throughout the entire thickness of the gray matter, overlap with each other and parallel that of their functional cohorts, the pyramidal neurons (Chapter 3). These functional territories are also sequentially stratified from older, lower, and larger ones to younger, smaller, and superficial ones. The appearance of the double-bouquet inhibitory interneurons and that of their dendritic and axonic profiles and cylindrical functional territories remains unchanged from any angle of view.

6.4 The Pyramidal-Chandelier System

The chandelier (axo-axonic) neuron is the smallest of all known inhibitory interneurons. They are quite common in the human motor and visual cortex (Marín-Padilla 1987). In one of the cases studied (an 8-month-old child), chandelier neurons were quite abundant in the visual cortex striate (area 17) and parastriate (area 18) regions, being particularly numerous at their transitional zone (Fig. 6.12d). This particular location for chandelier cells has been also mentioned of the visual cortex of rats (Somogyi 1977; Peters 1984), cats (Somogyi 1979; Fairen and Valverde 1980), rabbits (Müller-Paschinger et al. 1983) and monkeys (Somogyi et al. 1982). Although these interneurons seem to be particularly abundant at the visual cortex; they are also found in other cortical regions including the newborn motor cortex (Fig. 6.12a). There are no reasons why they should not be also present in other cortical regions that remain essentially unexplored.

The chandelier neuron is a small multipolar cell with round or globular body (ca. 20 μm diameter), several (six–nine) ascending and descending dendrites with a reduced lateral extension and by an axon with specific axo-axonic terminals (Fig. 6.12a–c). The dendrites are thin, irregular, have focal dilatations and a few long spines-like excrescences (Fig. 6.12d). They also have a "wavy" appearance and short "recurring" collaterals (Fig. 6.12d). The dendritic arbor covers a smaller functional territory than the axon (Fig. 6.12b–c).

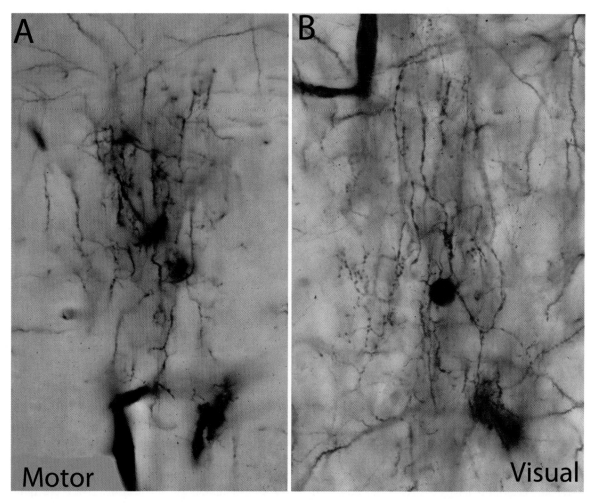

Fig. 6.12 Composite figure of photomicrographs, from rapid Golgi preparations of newborn infants motor (**a**) and visual cortex (**b**, **c**), illustrating the small size and concentrate dendritic and axonic profiles of chandelier inhibitory interneurons. Also illustrated is a camera lucida drawing (**d**) illustrating, in the human primary visual cortex (intersection of areas 17 and 18), the number, size, location, dendritic and axonic profiles, and the small rectangular functional territory of these inhibitory cells. These interneurons axo-axonic terminals form specific and short terminals arrays of axo-axonic synaptic contacts (candles) with the pyramidal neurons axon first segment. In Golgi preparation, in order to visualize these inhibitory interneurons specific axo-axonic terminals (candles) their pyramidal neuron functional cohort must be unstained (**a–d**)

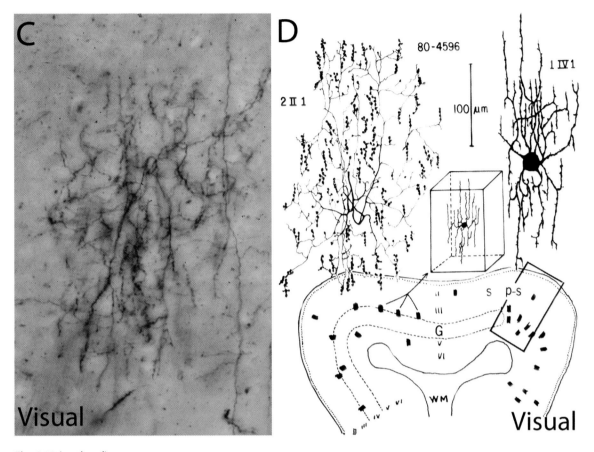

Fig. 6.12 (continued)

The most distinguishing feature of the neuron is its specialized axo-axonic terminals, referred to as candles, and hence its colorful nomination of chandelier cells (Figs. 6.13a–d, asterisks). The axon emerges from the top or bottom poles of the neuron body and immediately bifurcates into several ascending and descending collateral (Fig. 6.12c, d). From these collaterals emerge numerous ascending and descending short vertical branches that carry the neuron specialized synaptic terminals (Fig. 6.13a–d). These neurons specialized axonic terminals are unique structures characterized by short vertical arrays of presynaptic buttons or "candles" (Fig. 6.13, asterisks). In rapid Golgi preparation, these specialized axo-axonic terminals (candles) can only be visualized if the contacted and functional cohort pyramidal neurons are unstained. If the pyramidal neuron is stained, these fine axo-axonic terminals will not be adequately visualized. The structural complexity of these axo-axonic "candles" terminals depends on the number of contacts and could range from simple to complex ones. Simple ones are composed of a few axo-axonic "en-passant" contacts united by a very fine axonic fiber. The complex ones are longer, often have double arrays of presynaptic contacts interconnected by fine fibers (Figs. 6.13a–d, asterisks). In complex terminals, the number of axo-axonic buttons could ranges up to 10.

These inhibitory interneurons terminal axo-axonic arrays (candles) are essentially identical in all mammals so far investigated (Valverde 1983; Marín-Padilla 1987). However, the numbers of "candles" per neuron varies significantly among different mammals. In the human visual cortex, the number of specialized terminals ranges from 60 to 80 "candles." This low number of presynaptic contacts and the neuron's size will make the human chandelier cell the smaller among all mammalian species studied (Lund et al. 1979; Lund 1981; Valverde 1983; Marín-Padilla 1987).

Within the thickness of a rapid Golgi preparation (up to 250 μm) it is possible to visualize the chandelier cells in toto, including their complete dendritic and axonic profiles and their overall functional territory (Fig. 6.12a, b, d). The neuron occupies the center of a relatively small rectangular functional territory, which measures, roughly, 300 × 200 × 100 μm (Fig. 6.12d). Their roughly rectangular functional territory remains unchanged in both perpendicular and parallel cut rapid Golgi preparations.

Although, the human chandelier cell size and functional territory seems to be comparable to that of the monkey (Lund et al. 1979; Lund 1981; Valverde 1983), the neuron size and functional territory within the much larger human brain should be comparatively smaller than of the monkey. Moreover, the overall size of chandelier neuron's functional territories seems to decrease progressively in the course of mammalian phylogeny (Valverde 1983). It appears that the size and functional territory of these inhibitory interneurons decreases progressively through the hedgehog, mouse, rabbit, cat, and monkey (Valverde 1983). A possible interpretation of this phenomenon could be that, in the course of mammalian phylogeny, chandelier cells increase their functional specialization by reducing both the number of pyramidal neurons contacted and the size of their functional territory. During mammalian phylogeny, this type of inhibitory interneurons

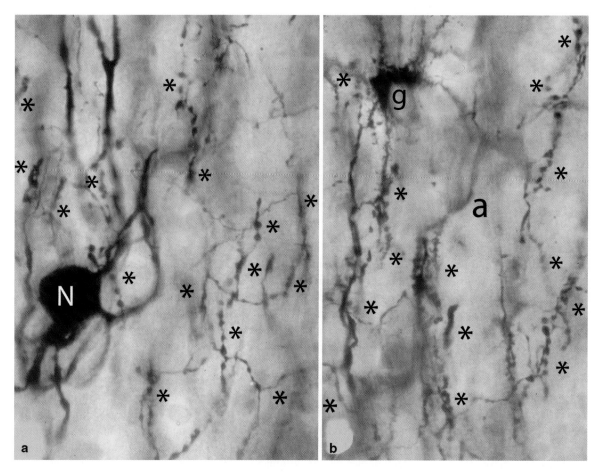

Fig. 6.13 Montage of photomicrographs (**a–d**), from rapid Golgi preparation of newborn's primary visual cortex, illustrating, at a higher magnification, the morphological features of the specific axo-axonic terminals (candles) of chandelier inhibitory interneurons. In these preparations, to visualize these specific terminals (*) the pyramidal neurons (functional cohorts) must be unstained. The composition, size, distribution, and number of synaptic axo-axonic contacts (*) of chandelier cells axon (**a**) are clearly demonstrated in these (**a–d**) illustrations. A chandelier cell small and globular body and its smooth wavy dendrites are also illustrated (**a**)

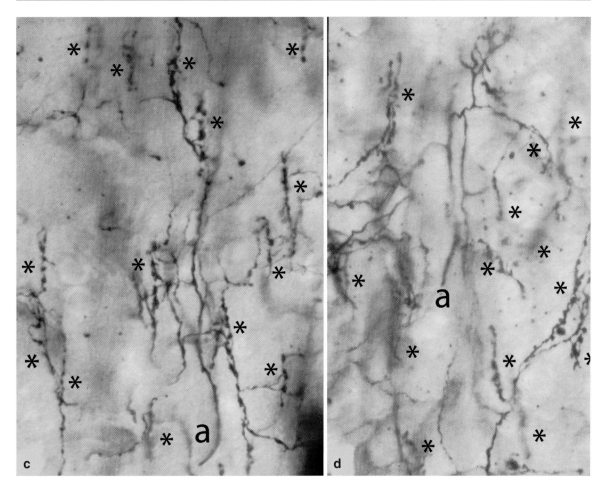

Fig. 6.13 (continued)

shows a tendency toward a reduction of its dendritic and axonic processes. A concentrated pattern of dendrites and axonic distribution has been considered to connote functional specialization (Ramón-Molliner and Nauta 1966; Szentágothai and Arbib 1974).

A functional parameter that supports the idea that chandelier cells might undergo a progressive specialization is the reduced number of axo-axonic terminals (candles) per neuron observed among different mammals. The number of axo-axonic contacts "candles," (or the number of pyramidal neurons contacted) reported in the literature decreases from 220–260 "candles" in rabbits, to 150–250 in cats, to around 100 in monkeys, to 60–80 in the humans visual cortex, per chandelier cell (Tömböl 1976, 1978; Somogyi et al. 1982; Müller-Paschinger et al. 1983; Valverde 1983; Marín-Padilla 1987). Moreover, the size of these inhibitory neuron's functional territory also decreases proportionally. A possible functional specialization of chandelier cells could be related to the increasing visual acuity and motor dexterity that characterize the evolution of mammals.

The functional territories of inhibitory neurons become delineated by the extension of its axonic terminals and hence the number of contacted pyramidal neurons within (Fig. 6.14). The functional territory of a basket cell is a large rectangular slab perpendicular to the pia and the long axis of the precentral gyrus (Fig. 6.14b); that of a double-bouquet interneuron is a vertical and cylindrical one (Fig. 6.14d–t); and, that of a chandelier cell is a small rectangular one (Fig. 6.14c). The number of pyramidal neurons (functional cohorts) contacted by a single basket cell of P1 stratum is probably a hundred (Figs. 6.2, 6.3, 6.5, and 6.14b). The apical dendrites of pyramidal neurons contacted within a single double-bouquet inhibitory interneuron functional

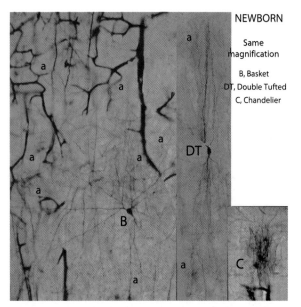

Fig. 6.14 This montage includes three photomicrographs of equal magnification, from rapid Golgi preparations of motor and visual cortex of newborn infants, illustrates comparatively the neuron size and extent of its corresponding functional territory of three inhibitory interneurons, namely: a deep basket (B), a double-tufted (DT), and a chandelier (C) cell. It is important to recognize and to appreciate the significant morphological, functional differences, and spatial orientation among some of the inhibitory interneurons of the human cerebral cortex. The functional implications of these differences are of paramount importance for understanding the function of the human brain

territory may range between from 6 to10 on the upper strata and from 12 to 16 on the lower ones (Figs. 6.10, 6.11, and 6.14d–t). The functional territory of a single chandelier cell of the human visual cortex could establish synaptic contacts with 60–80 pyramidal cells axons (Figs. 6.12, 6.13, and 6.14c). A Martinotti cell may establish synaptic contacts with the terminals dendritic bouquets of several pyramidal neurons of its strata. Possibly, a single Martinotti inhibitory cell may contacts with between 6 and 12 pyramidal neurons.

These significant differences are important contributing factors in the final functional output of the cerebral cortex pyramidal cells. It should be emphasized that the incorporation and functional maturation of these excitatory–inhibitory systems follows an ascending and stratified sequence that parallel that of their functional cohorts, the pyramidal neurons. While the developmental maturation of these inhibitory systems is an ascending process, their functional inputs may follow a descending pathway, from upper to lower strata, paralleling the pyramidal neurons descending functional network (Chapter 4). The motor capabilities and proficiency of each mammalian species will depend on the number, and extent of the excitatory–inhibitory neuronal systems established in their corresponding motor cortex.

The remarkable morphological (different sizes), functional (extend of functional territories), and spatial orientation differences among the inhibitory interneurons of the cerebral cortex are important observations (Fig. 6.14) which must be considered for the understanding and interpretation of the functions of the human brain.

It should be pointed out that besides the four excitatory–inhibitory systems described herein, there are additional ones throughout the cerebral cortex, targeting the pyramidal neurons of the various strata, which remain essentially unidentified. These additional systems should also be investigated, hopefully using the rapid Golgi procedure.

References

Andersen P, Eccles JC, Løyning Y (1963a) Recurrent inhibition in the hippocampus with identification of the inhibitory cells and its synapses. Nature 198:540–542

Andersen P, Eccles JC, Voorhoeve PE (1963b) Inhibitory synapses on somas of purkinje cells in the cerebellum. Nature 199:655–665

Blakemore C, Tobin EA (1972) Lateral inhibition between orientation detectors in the cat's visual cortex. Exp Brain Res 15:439–440

Cajal SyR (1893) Estructura asta de Ammon y fascia dentate. Anales de la Sociedad Española de Historia Natural 22:30–46

Cajal SyR (1899) Estudios sobre la corteza cerebral humana Estructura de la corteza motriz del hombre y mamíferos. Revista Trimestral Micrográfica 4:117–200

Cajal SyR (1911) Histologie du systéme nerveux de l'homme et des vertébrés, vol 2. Maloine, Paris

Eccles JC (1964) The physiology of the synapses. Academic, New York

Fairen A, Valverde F (1980) A special type of neuron in the visual cortex of the cat: a Golgi and electron microscopic study of chandelier cells. J Comp Neurol 194:761–779

Hubel DH, Wiesel TN (1977) Functional architecture of macaque monkey visual cortex. Proc R Soc Lond Ser B 189:1–59

Jones EG (1984) Laminar distribution of cortical efferent cells. In: Peters A, Jones EG (eds) Cerebral cortex, vol 1. Plenum Press, New York, pp 521–553

Jones EG, Hendry SH (1984) Basket cells. In: Peters A, Jones EG (eds) Cerebral cortex, vol 1. Plenum Press, New York, pp 309–336

Lewis DAT, Hashimoto T, Volk DW (2005) Cortical inhibitory neurons and Schizophrenia. Nat Rev Neurosci 6:312–324

References

Lund JS (1981) ntrinsic organization of the primate visual cortex, area 17 as seem in Golgi preparations. In: Schmitt FO, Worden FG, Adelman G, Dennis SG (eds) Organization of the cerebral cortex. MIT Press, Cambridge, MA, pp 105–124

Lund JS, Henry GH, Mac Queen CI, Harvey AR (1979) Anatomical organization of the primate visual cortex (area 17) of the cat: A comparison with area 17 of the macaque monkey. J Comp Neurol 148:599–618

Marín-Padilla M (1969) Origin of the pericellular baskets of the pyramidal neurons of the human motor cortex. Brain Res 14:633–646

Marín-Padilla M (1970) Prenatal and early postnatal ontogenesis of the human motor cortex. A Golgi study. II. The basket pyramidal system. Brain Res 23:185–191

Marín-Padilla M (1972) Double origin of the pericellular baskets of pyramidal neurons of the human motor cortex. A Golgi study. Brain Res 38:1–12

Marín-Padilla M (1974) Three-dimensional reconstruction of the pericellular baskets of the motor (area 4) and visual (area 17) areas of the human cerebral cortex. Zeitschrift für Anatomie und Entwicklungeschichte 144:123–135

Marín-Padilla M (1984) Neurons of layer I. A developmental study. In: Peters A, Jones EG (eds) Cerebral cortex, vol 1. Plenum Press, New York, pp 447–478

Marín-Padilla M (1987) The chandelier cell of the human visual cortex. A Golgi study. J Comp Neurol 265:61–70

Marín-Padilla M (1990) The pyramidal cell and its local-circuit interneurons: a hypothetical unit of the mammalian cerebral cortex. J Cognitive Neurosci 2:180–194

Marín-Padilla M, Stibitz G (1974) Three-dimensional reconstruction of the baskets of the human motor cortex. Brain Res 70:511–514

Müller-Paschinger IB, Tömböl T, Petsche H (1983) Chandelier neurons within the rabbit cerebral cortex. A Golgi study. Anat Embryol 166:149–154

Peters A (1984) Chandelier cells. In: Peters A, Jones EG (eds) Cerebral cortex, vol 1. Plenum Press, New York, pp 361–380

Ramón-Molliner R, Nauta WJH (1966) The idiodendritic core of the brain stem neurons. J Comp Neurol 126:311–335

Somogyi P (1977) A specific 'axo-axonic' interneuron in the visual cortex of the rat. Brain Res 136:345–350

Somogyi P (1979) An interneuron making synapses especially on the axon initial segment of pyramidal neurons in the cerebral cortex of the cat. J Physiol (Lond) 296:18–19

Somogyi P, Freund TF, Cowey A (1982) The axo-axonic interneuron in the cerebral cortex of the cat, rat, and monkey. Neurosciences 7:2577–2609

Somogyi P, Cowey A (1984) Double bouquet cells. In: Peters A, Jones GE (eds) Cerebral cortex, vol 1. Plenum Press, New York, pp 337–358

Szentágothai J (1965) The synapses of short local neurons in the cerebral cortex. Symp Biol Hungary 5:251–276

Szentágothai J (1975) The "module concept" in the cerebral cortex architecture. Brain Res 95:475–496

Szentágothai J (1978) The neuron network of the cerebral cortex: a functional interpretation. Proc R Soc Lond Ser B 201:219–248

Szentágothai J, Arbib MA (1974) Conceptual models of neural organization. Neurosci Res Prog Bull 12:383–286

Tömböl T (1976) Golgi analysis of the internal layers (V–VI) of the cat visual cortex. Exp Brain Res 1:292–295

Tömböl T (1978) Comparative data on Golgi architecture of interneurons of different cortical areas in the cat and rabbit. In: Brazier MAB, Petsche H (eds) Architectonics of the cerebral cortex. Raven Press, New York, pp 59–76

Valverde F (1983) A comparative approach to neocortical organization based on the study of the brain of the hedgehog (Erinaceous europeus). In: Grisolia S, Guerry CC, Samson F, Norton S, Reinoso-Suarez F (eds) Ramón y Cajal contribution to neurosciences. Elsevier, Amsterdam, pp 149–170

Human Cerebral Cortex Intrinsic Microvascular System: Development and Cytoarchitecture

The rapid Golgi procedure is also an excellent tool to study the development of the intrinsic microvascular system of the mammalian cerebral cortex and, in general, of the central nervous system (CNS). Sprouting of capillaries and the sequential evolution of short-linked anastomotic plexuses are stained and clearly visualized with this procedure (Marín-Padilla 1985, 1987, 1988, 1997, 1999). The formation of short-linked anastomotic capillary plexuses and their subsequent remodeling and adaptations represent the basic elements of the cerebral cortex microvascular system. The embryonic mammalian cerebral cortex, as a neuroectodermal tissue, is originally deprived of an intrinsic microvascular system and will need to build one. The development of the cerebral cortex microvascular system is a complex processes characterized by the sequential evolution of three well-defined compartments. The sequential evolution of these three vascular compartments is explored herein, using both the rapid Golgi procedure and the electron microscope.

The mammalian cerebral cortex vascular system is characterized by the sequential development of the meningeal or pericortical, the Wichows–Robin or extrinsic and the intracortical or intrinsic compartments, respectively (Marín-Padilla 1987). While the development of the first two compartments is extrinsic to the nervous tissue proper, the third one constitutes its intrinsic system, with capillaries growing in close proximity to the nervous elements. By active capillary angiogenesis and reabsorption and by sequential remodeling, each compartment undergoes progressive readaptations to the developing cerebral cortex external and internal changing configurations. Understanding the sequential developmental and interrelationships among these compartments is essential. The sequential development of the neocortex extracortical, extrinsic and intrinsic vascular compartments is described below.

The embryo vasculogenesis starts "in situ" from angioblastic cell islands scattered through the mesoderm, yolk sac, and body stalk. Some of these islands fused and undergo a process of internal liquefaction resulting in their early canalization (1920a, b). The mechanism of this early canalization remains poorly understood (Manasek 1971). Following their canalization, the angioblastic islands establish interconnections among themselves forming a vascular plexus composed of irregular primordial blood vessels. These primordial vessels are surrounded by incomplete basal lamina, growths by endothelial cells division and sprouting and, eventually, establish a precirculatory plexus throughout the embryo. The blood cells within these primordial vessels seem to evolve from the original angioblastic cells. Eventually, the embryo precirculatory plexuses establish connections with the heart, blood starts to circulate, and its early arterial and venous systems evolve progressively.

The CNS vascularization begins at the myelencephalon and ascended through the metencephalon, mesencephalon, diencephalon, and striatum and, finally, reaches the telencephalic vesicle (Mall 1904; Streeter 1918; Strong 1964; Bär and Wolff 1972; Kaplan and Ford 1976; Folkman 1982; Marín-Padilla 1985; O'Rahilly and Muller 1986;). This cephalic-vascular progression coincides with the CNS ascending cytoarchitectural and functional maturations. The cephalic vascular plexus is the last one formed. It starts its formation as early as the 4th week of gestation (Streeter 1918; Padget 1948, 1957; Hamilton et al. 1972) and, by the 7th week, is the most prominent vascular plexus in the embryo (Fig. 7.1a). It envelops the entire developing brain and eventually establishes its meningeal vascular compartment. Most of the main arteries, veins, and venous sinuses of the adult brain meningeal system are already recognizable by the 7th week of gestation (Fig. 7.1a). At this age, an extensive pial anastomotic capillary plexus is formed covering the embryonic neocortex surface, which supplies oxygen for subpial primordial fibrillar and neuronal systems. At this age, the cerebral cortex is still avascular (Chaps. 2 and 3).

At this age, the brain meningeal compartment has three distinct and interconnected vascular strata (Marín-Padilla 1987). The outer or dural stratum has large vessels from which the meningeal venous sinuses will evolve (Fig. 7.1a, b). The intermediate or arachnoidal stratum contains the vessels from which the main cephalic arteries and vein will evolve (Fig. 7.1b). The arachnoidal arterial and venous systems undergo profound developmental transformations, as they adapt to the developing brain expanding and changing configuration. These arterial and venous transformations have been elegantly described and illustrated by Padget (1948, 1957). Through her illustrations is it possible to follow the progression of the main meningeal arteries and/or veins throughout the human brain prenatal development. However, there is no mention in her embryological studies of the meninges third vascular compartment, namely: the pial anastomotic capillary plexus. Moreover, there are very few descriptions of the development of this important vascular compartment (Krahn 1982; Morse and Low 1972). Two reasons might explain these lapses. First, the pial capillaries are invisible to the naked eye and, second, this plexus is invariably removed together with the meninges, leaving the brain surface apparently without vessels. However, the pial capillary plexus is an important component of the meningeal inner vascular

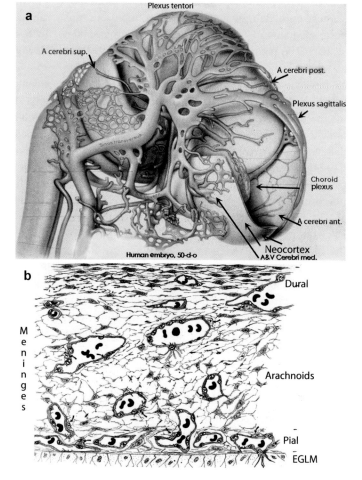

Fig. 7.1 Composite figure including a color drawing (**a**) of the intracranial cephalic vasculature of a 50-day-old fetus showing its basic arterial and venous systems and its venous plexuses and a camera lucida drawing (**b**) from rapid Golgi preparations showing the composition of the meninges around the embryonic cerebral cortex, at this age. Most of the basic components of the cephalic meningeal vasculature of the adult brain (A) are already recognizable at this embryonic age. The cerebral vesicle has been opened (A) to show the choroid plexus and the thin and still avascular cortical mantle. The cerebral cortex pial capillary plexus is not shown. (**b**) This drawing illustrates meninges dural, arachnoidal, and pial cellular and vascular components developing cerebral cortex of a 50-day-old human fetus. It also illustrates the close proximity of pial anastomotic capillary plexus to the cortex external glial limiting membrane (EGLM). A pial capillary that has established contact, by advancing endothelial cell filopodia, with the cortex EGLM and another one has already perforated through it. The color illustration depicted in A is a modified reproduction from Streeter's original drawing (Streeter 1918)

compartment and will play a crucial role in the development of the cerebral cortex microvascular system (Fig. 7.1b).

By the 7th week of gestation, the meninges around the brain are about 100 μm thick and have three distinct lamellae (Fig. 7.1b). Its outer or dural lamella, formed immediately under the neurocranium, is composed of closely arranged and elongated meningeal cells and venous sinuses (Fig. 7.1b). Its intermediate or arachnoidal is the thicker and best vascularized lamellae. It has numerous blood vessels of different calibers and its cells are loosely arranged with large intercellular spaces between them. At this age, its vessels have already started to differentiate into arterioles or venules (Fig. 7.1a). The pial or inner lamella is composed of a short-linked anastomotic capillary plexus that extends through the entire surface of the brain (Fig. 7.1b). Among its capillaries there are elongated pial cells and collagen fibers. The dural, arachnoidal, and pial vessels constitute the cerebral cortex meningeal or pericortical vascular compartment. Eventually, growing capillary sprouts from the pial plexus perforate the cortex external glia limiting membrane (EGLM), penetrate into the nervous tissue, and participate in the formation of its microvascular system (Fig. 7.1b). While the prenatal development of the meningeal dural and arachnoidal lamella vascular systems have been well documented, that of the inner or pial lamellae has not; its embryonic maturation is explored below.

Pial Capillary Plexus. The human cerebral cortex intrinsic microvascular system starts to develop around the 8th week of gestation. Consequently, its early embryonic (Neuroectodermal, Marginal Zone, Primordial Plexiform, and early Pyramidal Cell Plate) stages evolve within a still avascular neocortex (Chap. 2). The close proximity of the embryonic elements of the embryonic cerebral cortex to the pial anastomotic capillary plexus allows sufficient oxygen diffusion to permit their evolution.

The pial capillaries are interconnected with the arachnoidal vessels and constitute an extensive anastomotic capillary plexus that covers the entire developing brain. This pial capillary plexus will persist throughout the cerebral cortex prenatal and postnatal developments. The pial capillaries' size varies ranging from small ones built by single endothelial cells (probably newly formed and/or growing ones) to larger ones composed of several cells joined by tight junctions (Figs. 7.2a, arrows). They form a short-linked anastomotic capillary plexus that extends, in close proximity to the neocortex external glial limiting membrane (EGLM) (Figs. 7.2a and 7.3a). The pial capillaries, by active angiogenesis and reabsorption, adapt to the expanding cerebral cortex and will the source of all the perforating vessels that penetrate into it. Eventually, the pial capillary plexus will cover the entire surface of the neocortex including its sulci and gyri. During the prenatal and postnatal cortical maturations, perforating pial vessels continues to participate in the cerebral cortex expanding intrinsic microvascular system (Pape and Wigglesworth 1979; Duvernoy et al. 1981; Marín-Padilla 1988).

The pial capillaries are separated by elongated pial cells, a few collagen fibers and by tissue spaces (Figs. 7.2a and 7.3a). The pial cells are considered to be meningeal elements that share features with fibroblasts (collage production), mesodermal cells (phagocytosis), and epithelial cells (epithelial-like lamellae formation) (Andres 1976a, b, Krisch et al. 1982, 1983; Krisch and Buchheim 1984). The pial capillaries are surrounded by basal lamina material manufactured by the endothelial cells. The pial capillaries are also surrounded by perivascular spaces, which play a role in the cerebrospinal fluid circulation and could represent the port of entry for the meningeal perivascular lymphatics (Jones 1970; Morse and Low 1972; Pessacq and Reissenweber 1972; Nabeshina, Reese, and Landis 1975; Andres 1976a, b; Krisch, Leonhardt and Oksche 1982, 1983; Oda and Nakanishi 1984; Krisch and Buchheim 1984).

Active angiogenesis and formation of capillary sprouts are quite common among the pial capillaries. Growing capillaries are characterized by leading endothelial cells with considerable membrane activity (Fig. 7.2a). The growing capillaries' leading endothelial cells develop filopodia projecting both inside and outside the capillary lumen and some of them start to make contacts with the neocortex EGLM (Figs. 7.2a and 7.3a). In electron micrographs, the growing capillary leading endothelial cells are characterized by dark cytoplasm with abundant granular endoplasmic reticulum filled with dense granular proteinaceous material (Figs. 7.2a–d and 7.4a–c). This material may be used in manufacturing the growing vessel basal lamina and may participate in the formation of its first lumen (Manasek 1971). Endothelial cell mitoses are also frequent in growing capillaries.

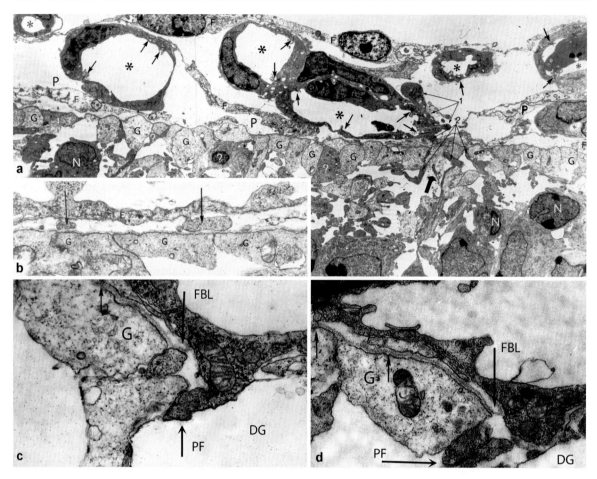

Fig. 7.2 Montage of photomicrographs, from electron microscopic studies of cerebral cortex early vascularization from 12-day-old hamster embryos, showing the contact of some pial capillaries (**a**) with the cortex EGLM, the direct (**b**) endothelial filopodia contacts with the fusion of vascular and glial basal laminae, the EGLM filopodia perforation and penetration (**a, c, d**) into the nervous tissue, and the fusion of vascular and glial basal laminae at the filopodia entrance site with the formation of funnel that remains open to the meningeal intercellular spaces. Also illustrated are the fused glial endfeet (G) that constitute the cortex EGLM, several pial capillaries (*), and some subpial (N) neurons

Fig. 7.3 Composite figure of camera lucida drawings (**a, b**), from rapid Golgi preparations of 12-day-old hamster embryo's cerebral cortex, illustrating: the three (endothelial cell contact, filopodia perforation, and capillary perforation) stages of the CNS vascular perforation; an electron photomicrograph (**c**) a Virchow-Robin Compartment (V-RC) (from Jones 1970), and a glial fibrillary acidic protein stain preparation (**d**) of a V-RC from an adult human brain. The V-RC represents a perivascular space formed, at the perforating vessel entrance (**b**), between the glial and vascular basal laminae. In the adult brain, the V-RC perforating vessel (**c**) is surrounded by a space established between the vascular and the glial basal lamina. Some pericytes (**c**) are transformed into smooth muscle cells that become incorporated into the vascular basal lamina. While the V-RC inner wall is walled by the vascular basal lamina, its outer wall is lined by the glial basal lamina (**c, d**). The V-RC and may contain inflammatory (macrophages) cells (**c, d**). Inflammatory cells increase significantly in the V-RC in CNS infections. The V-RC represents a pre-lymphatic draining system of the CNS that communicates directly with meningeal perivascular lymphatics. GFAP preparation demonstrates the V-RC glial outer wall and some macrophages. Key to abbreviations: M = macrophages; BV = blood vessels; P = pericytes; BC = blood capillary; PC = perforating vessel; EMG = electron micrograph; GFAP = glial fibrillary acidic protein stain; V-RC = Virchow–Robin Compartment; EGLM = external glia limiting membrane

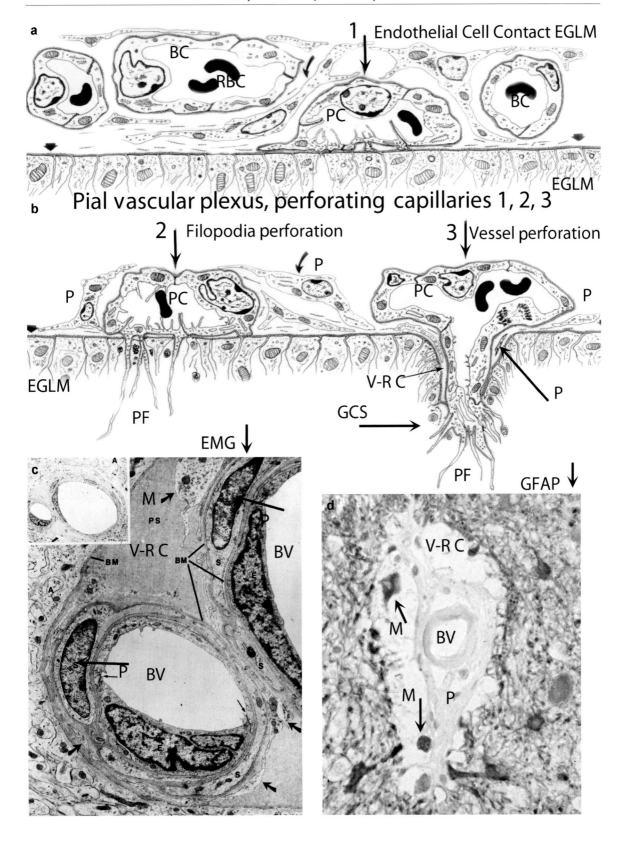

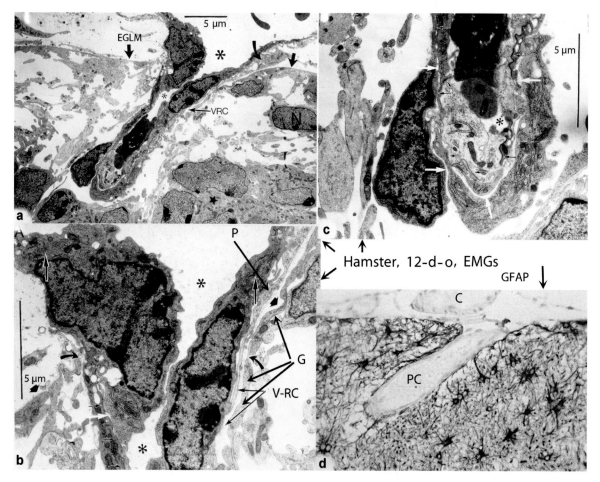

Fig. 7.4 Montage of photomicrographs (**a–c**), from electron microscopic studies of 12-day-old hamster embryos, showing the CNS entrance site of a perforating vessel (**a**) surrounded by the V-RC (**a, b**) and its growing tip (**c**) of sliding endothelial cells. Also illustrated (**d**) is a glia fibrillary acidic protein (GFAP) preparation showing a perforating vessel within a V-RC with a GFAP positive outer glial wall composed of adjoining glial endfeet processes. Many of the contributing glial cells are also stained. The V-RC outer glial wall is a continuation of the cortex superficial EGLM (**a–d**). The perforating capillary growing tip (**c**) is composed of several sliding endothelial cells deprived of basal lamina material and separated by several intercellular spaces (white arrows). Key to abbreviations: EMGs = electron micrographs; GFAP = glia fibrillary acidic protein; * = capillary lumen; V-RC = Virchow–Robin Compartment; EGLM = external glial limiting membrane; G = glial cells (astrocytes). Scales = 5 μm

During the vascular perforation of the developing cerebral cortex, the EGLM and the leading tip of growing capillaries establish contact that precedes the capillary perforation. The growing tips of pial capillaries approach and establish direct contacts with the neocortex EGLM and their corresponding basal lamina fussed (Figs. 7.2a–d and 7.3a, b). The EGLM is composed of closely apposed radial glia endfeet that joined by tight junction cover the entire cortical surface and manufacture its basal lamina. The EGLM constitutes the neocortex (and entire CNS) anatomical and functional barrier. It must be maintained intact through neocortical development by the continued incorporation of new glial endfeet. This barrier is only perforated by entering capillaries and nerve terminals and by exiting nerve terminals (Andres 1976a, b; Marín-Padilla 1985). The vascular perforation of the embryonic neocortex EGLM has been studied using both light and electron microscopic preparations (Krahn 1982; Marín-Padilla 1985, 1987, 1988).

The neocortex EGLM vascular perforation is characterized by the sequential progression of four interrelated processes, namely: (a) capillary approach and filopodial contact with the cortex EGLM; (b) endothelial filopodia perforation of the EGLM and penetration into the nervous tissue; (c) capillary penetration into nervous tissue with formation of a perivascular space around it; and, (d) subsequent establishment of the neocortex intrinsic microvascular system (Figs. 7.2–7.5). The perivascular space will maintain the perforating vessel extrinsic to the nervous tissue and will remain open to the meningeal compartment (Figs. 7.2–7.5).

Pial Capillary Approach, Contact & Filopodia Perforation. Growing pial capillaries approach the neocortex EGLM and establish direct filopodial contact with its basal lamina (Fig. 7.2a, b). Both the approaching capillary and the EGLM basal laminae establish contact and subsequently fuse around the perforation filopodium (Fig. 7.2b–d and 7.3a, b). The advancing endothelial filopodium dissolves the basal laminae, liquefies the glial endfeet junction, and penetrates into the nervous tissue (Fig. 7.2c, d and 7.4a, b). At the filopodium entrance site, glial and vascular basal laminae refused, forming between them, a funnel, which remains open to pial spaces and will, accompanies the perforating capillary into the neocortex (Fig. 7.2c, d).

Capillary Perforation & Virchow–Robin Compartment. Eventually, the perforating vessel penetrates into the cortex surrounded by a perivascular space (Fig. 7.3b, c and 7.4a–d). The surface EGLM seems to accompany the perforating vessel as new glial endfeet are progressively incorporated, keeping the perforating capillary extrinsic to the nervous tissue (Figs. 7.3b, c and 7.4a–d). This perivascular space, known as the Virchow–Robin Compartment (V-RC), accompanies the perforating vessels throughout its entire length (Jones 1970). The V-RC is formed between the inner basal lamina of the perforating vessel and the outer basal laminae of glial origin (Figs. 7.3c, d and 7.4a–d). The V-RC remains open to pial and arachnoidal spaces (Figs. 7.3b–d and 7.4d). The perforating vessel is accompanied by pial cells (pericytes), which remain within the perivascular V-RC (Figs. 7.3c, d and 7.4b). Eventually, some pericytes become incorporated into the vascular basal lamina and are transformed into smooth muscle cells (Jones 1970). As the perforating capillary advances into the neocortex, its perivascular V-RC advances with it by the progressive incorporation of new glial endfeet. The perforating vessel leading endothelial cells remain outside of the perivascular space and its advancing filopodia, surrounded by an incomplete basal lamina, grow among the nervous elements (Figs. 7.3b and 7.4a, c).

During early embryonic development, the perforating vessels reach, unbranched, the cortex paraventricular zone remaining inside the V-RC and, hence, extrinsic to the nervous tissue (Fig. 7.4a). The perforating vessels within the V-RC represent the extrinsic microvascular system of the cerebral cortex. Eventually, these early perforating vessels intercommunicate among themselves establishing an anastomotic capillary plexus that extends throughout the neocortex paraventricular zone. This is the first of numerous additional anastomotic capillary plexuses, between adjacent perforating vessels, which will form during the cerebral cortex ascending maturation (Fig. 7.5a). As the neocortex increases in thickness, the V-RC length also increases by the continued incorporation of new glia endfeet. Both, electron micrographs and glial fibrillary acidic protein stain, demonstrate the V-RC composition and organization (Figs. 7.3c, d and 7.4a, d). It also remains extrinsic to the nervous tissue. In the adult cerebral cortex, the perforating vessel continues to be V-RC central component, the vascular basal lamina, its inner wall, and the glial endfeet basal lamina, its outer wall (Fig. 7.3c, d). Smooth muscle cells become included within the vascular basal lamina (Fig. 7.3c). The V-RC perivascular space often contains inflammatory cells and phagocytic macrophages passing through or into the nervous tissue (Figs. 7.3c, d). In many pathological conditions, the number of inflammatory cells and of macrophages, within the V-RC, increases significantly. microvascular system.

Since the cerebral cortex, as well as the entire CNS, is deprived of a lymphatic system, the V-RC may function as a prelymphatic drainage system, in both physiological and pathological conditions. The V-RC perivascular space is the only access for inflammatory cells to move in and out of the nervous tissue, in both physiological and pathological situations. The V-RC central vessel is probably the only one capable of responding in any kind of inflammatory situation. Because the V-RC remains open, it communicates with the meningeal spaces and with its perivascular lymphatic system. This connection has been recently pointed out in several studies (Casley-Smith, Földi-Börcsök and Földi 1971; Pile-Spellman et al. 1984; Krisch and Buchheim 1984).

It is important to emphasize that during the cerebral cortex development, the distance between perforating vessels, and hence of V-RCs, remains constant and ranges from 400 to 600 µm (Fig. 7.5b). This distance

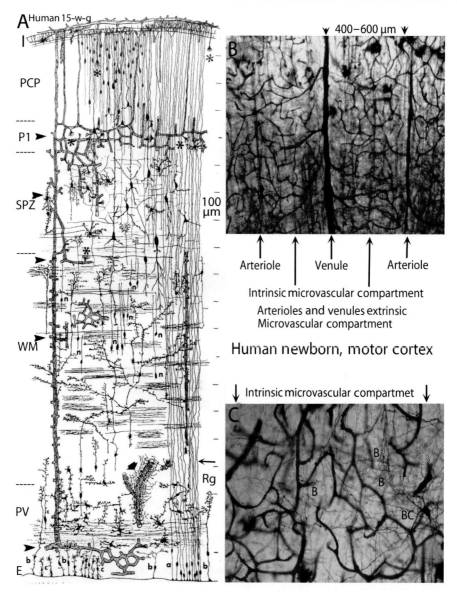

Fig. 7.5 Composite figure including a montage of camera lucida drawings (**a**), from rapid Golgi preparations of a 15-week-old human fetus and two color photomicrographs from rapid Golgi preparations illustrating some aspects (**b, c**) of the cerebral cortex extrinsic and intrinsic microvascular systems at this age. A. Reproduces a vertical section of the motor cortex measuring 800 μm wide, 2,000 μm high, and 150 μm thick illustrating the basic neuronal, fibrillar, microvascular, and glia cytoarchitecture of the human motor cortex at this age. Especially noticeable are the numerous capillary plexus already formed (*arrow heads*) between adjacent perforators and the newly established one in the lower gray matter coinciding with the maturation of the first pyramidal P1 cell stratum. The remaining features are self-explanatory (Chap. 3). (**b**) Reproduces a microscopic view of the cortex microvascular system, including an exiting venule surrounded by adjacent and equidistant entering arterioles (extrinsic vascular component) and the anastomotic capillary plexus established between them (intrinsic vascular component). The distance between entering arterioles and exciting venules (B) is uniforme throughout the entire cerebral cortex and ranges from 400 to 600 μm. C. Reproduces, at a high magnification, the 3-D organization of the anastomotic capillary plexus, formed between adjacent vessels, which represent the cortex intrinsic microvascular system as well as the so-called blood–brain barrier. Also illustrated are the intercapillary spaces and the presence of a basket cell (BC) as well as several pericellular baskets (B). Key to abbreviations: I = first lamina; PCP = pyramidal cell plate; P1 = first (P1) pyramidal cell functional stratum; SPZ = subplate zone; WM = white matter; PV = paraventricular zone; E = ependymal surface; Rg = radial glia; * = growing capillaries

that is nearly identical throughout the entire cerebral cortex will be maintained throughout both its prenatal and postnatal maturations. In my opinion, the similar distance between perforating vessels throughout the cerebral cortex represents a physiological necessity, which may reflect the overall distance that inflammatory cells could travel in and out of the nervous tissue. This distance also determines the size and extent of the cerebral cortex intrinsic microvascular system. Moreover, the capillaries of the cerebral cortex intrinsic microvascular system may be incapable of responding to any inflammatory event (see below).

Intrinsic Microvascular System, Development & Structure. During embryonic development, the perforating vessels advance from pial to subependymal zone. Their leading endothelial sprouts continue to grow, establishing capillary intercommunications between adjacent perforators. These early vascular interconnections establish anastomotic capillaries plexus between adjacent perforators, which extend throughout the developing cortex entire paraventricular zone. The interconnecting capillaries of this early plexus are deprived of perivascular space and are surrounded by a single basal lamina. Their single basal lamina could be the result of the fusion of vascular and glial laminae and/or only represents the glial basal lamina. This early subependymal capillary anastomotic plexus represents the first component of the cortex intrinsic microvascular system (Fig. 7.5a). New anastomotic capillary plexuses between adjacent perforators will continue to develop following an ascending progression that coincides and parallels the cerebral cortex ascending neuronal and functional maturations.

By the 15th week of gestation, the developing neocortex has intrinsic capillary plexuses between adjacent perforators, at many levels (Fig. 7.5a). At this age, the deep paraventricular capillary plexuses extend throughout the entire cortex. There are also additional intrinsic anastomotic capillary plexuses at various levels of the cellular matrix. This zone is quite active during the early embryonic development of the cerebral cortex. Additional anastomotic capillary plexuses, between adjacent perforators, are also recognized at various levels throughout the white matter and the subplate zones (Fig. 7.5a). At this age, the first gray matter anastomotic capillary plexus is starting to form throughout the pyramidal cell plate (PCP) lower region (Fig. 7.5a). Its establishment coincides with the starting functional maturation of the PCP deeper and older pyramidal neurons and the establishment of the first pyramidal cell P1 stratum (Fig. 7.5a). The formation of this first gray matter anastomotic capillary plexus will extend, between adjacent perforators, throughout the entire cortex following a ventral to dorsal gradient. This gradient coincides with that of the progressive maturation of the first pyramidal cell P1 stratum (Chapter 3). New anastomotic capillary plexuses, between the adjacent perforators, will continue to form throughout the gray matter paralleling its ascending cytoarchitectural and functional maturations. By the time of birth, intrinsic anastomotic capillary plexuses, between adjacent perforators, are recognized throughout the entire gray matter thickness, including those through pyramidal cell strata P1–P6 (Figs. 7.6a–d).

Therefore, the cerebral cortex intrinsic microvascular system is established between adjacent and equidistant perforating vessels through the cortex (Figs. 7.5a–c, 7.6a–d, and 7.7a, b). Growing capillaries emerge from the perforating vessel, cross the V-RC and reenter into the nervous tissue by a process analogous to the original perforation of the cerebral cortex EGLM. The growing capillary from the perforating vessel approaches and perforates the V-RC outer glial wall entering into the nervous tissue covered by a single basal lamina, possibly of glial origin. The advancing intrinsic capillary is progressively covered by the incorporation of new endfeet from the protoplasmic astrocytes of the gray matter. The new intrinsic capillaries establish contacts with those from adjacent perforating vessels establishing a short-linked 3-D anastomotic plexus between them (Figs. 7.5b, c, 7.6b, c and 7.7a, b). The intrinsic anastomotic capillary plexus, between adjacent perforators, represents the so-called blood–brain barrier and its capillaries are the only ones involved in the central nervous system functional activity.

A 3-dimensional short-link anastomotic plexus is progressively established between anterior, posterior, and laterals adjacent perforating vessels. The intrinsic capillary plexus 3-dimensional extension ranges between 400 and 600 µm in every direction (Fig. 7.6b–d). Within the anastomotic plexus, the length of the capillary loops varies considerably ranging between 60 and 110 µm (Fig. 7.6c). All intrinsic capillaries have uniform configuration and thickness (Fig. 7.6c). The diameter of the intercapillary spaces formed between these capillaries range from 100 to 150 µm (Figs. 7.5b, c, 7.6b, c, and 7.7a, b). These spaces are occupied by the gray matter functional elements (Fig. 7.5c). The

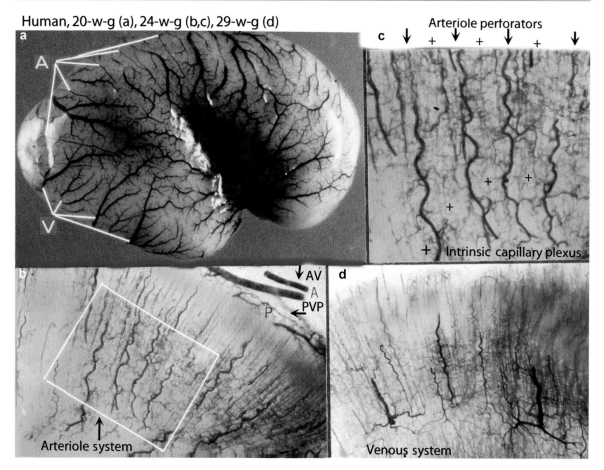

Fig. 7.6 Montage of photomicrographs showing (**a**) the meninges vasculature with main arteries and veins, from a 20-week-old human fetus and vertical cortical sections of color-injected arterial (**b**) and venous (**d**) systems of the human cortex and a high power view (**c**) of the equidistant perforating vessels with the intrinsic anastomotic capillary plexus (**+**) between them. The perforating vessels include both entering arterioles and an exiting venule. The pial capillary plexus of the meningeal vascular compartment is unrecognizable to the naked eye but it is clearly visible Pial Vascular Plexus vertical cortical sections (**b**). (**b–d**) are reproduced from Pape and Wigglesworth's original study (1979)

cortex intrinsic anastomotic capillary plexus is constantly renewing itself by both capillary angiogenesis and reabsorption, responding to the embryonic and/or adult cerebral cortex functional needs. The gray matter intrinsic capillary system has many more and shorter capillary loops than the one formed throughout the white matter. The white matter intrinsic capillary plexuses have fewer and longer capillaries and the intercapillary spaces are also larger (Figs. 7.6 and 7.7).

It is important to point out, that the pial capillaries perforate the cortex EGLM only to enter into the nervous tissue but not to exit it. This feature is also applicable to emerging capillaries from the V-RC vessels. Emerging capillaries from vessels within the V-RC perforated its outer glial wall to enter into the nervous tissue but not to return into it. Eventually, circulatory dynamic forces will determine which perforating vessels become entering arterioles and which one exiting venules. In the newborn cerebral cortex, each exiting venule is surrounded by 6–8 entering arterioles defining together a vascular territory (Fig. 7.5a–c, 7.6b–d, and 7.7a, b).

Throughout the brain surface, the vascular entrances of perforating vessels are quite small and nearly invisible to the naked eye. A magnifying glass will be needed to recognize the numerous small vascular orifices scattered throughout the entire brain surface. Therefore, it is not surprising that after the removal of the meninges, which invariably will carry the pial anastomotic capillary plexus, the brain surface should appears to be essentially deprived of blood vessels (Fig. 7.8a, b). A fact that could possibly explain why

Fig. 7.7 Composite figure showing, comparatively, the structure, composition, and organization of the cerebral cortex microvascular system from (**a**) a newborn infant (rapid Golgi preparations) and (**b**) from an adult brain (intravascular casting). The illustrations reproduce a vertical section of the cerebral cortex between a perforating arteriole that reaches the white matter and an exiting venule from the gray matter and include the intrinsic anastomotic capillary plexus formed between them. In both brains, the anastomotic capillary plexus of the gray matter is richer than that of the white matter. Perhaps, this vascular difference explains the greater vulnerability of the white matter to anoxia. Both the newborn and the adult cortical sections include the first lamina, the gray (GM), the subplate (SP) zone, and part of the white matter (WM). The remarkable anatomical and histological similarities between the infant and the adult microvascular systems are indeed extraordinary, which implies that they have remained essentially unchanged during the cerebral cortex developmental maturation, regardless of being totally renewed countless times. Various cortical strata are illustrated in both illustrations: first lamina (I) with Cajal-Retzius axon terminals (C-R at), the gray matter (GM), the subplate zone (SP), the white matter (WM), and the P1 pyramidal neuronal stratum. The intravascular casting of the adult cerebral cortex, is reproduced from Duvernoy, Delon, and Vannson's study (1981)

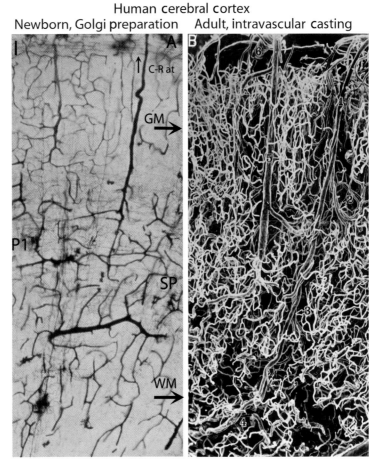

the pial capillary plexus, despite its significance, has received so little attention in the literature remaining practically ignored.

Throughout the cerebral cortex, the vascular orifices (V-RCs) of entering and/or exiting vessels are small, ranging between 2 and 3 mm in diameter and the distance between them remains essentially constant, ranging from 400 to 600 μm. Although there are more entering vessels than exiting ones, is not possible to recognize either. It is also important to emphasize that pial capillary perforations will continue to occur as long as the cerebral cortex continue to grow and expand. Also that the intervascular distance between perforators will remains essentially unchanged (Figs. 7.6 and 7.7). As the neocortex expands, numerous additional perforating vessels destined to its maturing gray matter, where most neurons reside, are progressively incorporated. Few of these new perforators will reach the white matter (Fig. 7.7a, b). The number of perforators that only reach the gray matter is considerably larger than those reaching the white matter. This is an important observation that may explain the greater vulnerability of the newborn infant white matter to hypoxia (Marín-Padilla 1997, 1999; Marín-Padilla et al. 2002). The disparity between the number perforators reaching the gray matter versus those reaching the white matter will continue to increase during the cerebral cortex postnatal maturation.

By mid-gestation, by the injection of different color dyes to arteries and veins, the various vascular compartments of the human brain can be clearly outlined and demonstrated (Fig. 7.6a–d). The meningeal compartment with its essential arteries and veins and its invisible pial capillary plexus can be readily appreciated (Fig. 7.7a). The entire meningeal compartment, including its pial plexus, can be removed leaving the cortical surface essentially avascular to the naked eye (Fig. 7.8a, b). Vertical sections of the injected brain demonstrate the arteriolar and venular nature of the

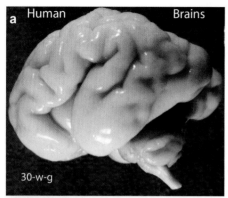

in the adult brain (Fig. 7.7 compare a and b). It should be emphasized that only the single basal lamina capillaries of the intrinsic capillary plexus, representing the blood–brain barrier, participate in the functional activity of both the developing and/or the mature brain.

The capillaries of developing cortex intrinsic microvascular system are very active by both angiogenesis and reabsorption. Rapid Golgi preparations are undoubtedly the best procedure to explore the growing (sprouting) as well as the reabsorbing capillaries throughout developing gray matter. In these preparations, the silver precipitates on endothelial cells (lipid membrane) of the vessel wall as well as on the advancing filopodia of growing capillaries (Figs. 7.9a, b and 7.10a–f). The sprouting capillary and its radiating filopodia are clearly

Fig. 7.8 Composite figure showing the smooth and apparently avascular surface of the cerebral cortex observed after the removal of the meninges, from a 30-week-old fetus (**a**) and from an adult brains (**b**). To view the innumerable, small, and equidistant vascular orifices throughout the brain surface (and that of other CNS regions), from entering arterioles and exiting venules, the use of a magnifying glass will be necessary. In general, the separation between these innumerable vascular orifices ranges between 400 and 600 μm. throughout the cortex

perforating vessels, their uniform intervascular distances, and the intrinsic capillary anastomotic plexus formed between contiguous perforators (Fig. 7.6b, c). Also the larger number of perforators reaching the gray matter is compared with the fewer ones reaching the white matter (Fig. 7.6b and 7.7a, b). At this age, the ratio of perforating vessels reaching the gray and with those reaching the white matters ranges around 8–2. This ratio will change, as additional perforating vessels destined to the gray matter will continue to increase during the brain postnatal maturation. The intrinsic sort-linked anastomotic capillary plexus formed between adjacent perforators throughout the gray matter is also clearly recognized (Fig. 7.6b, c). The overall structural organization of the intrinsic anastomotic capillary plexus between adjacent perforators of the newborn cerebral cortex remains essentially unchanged

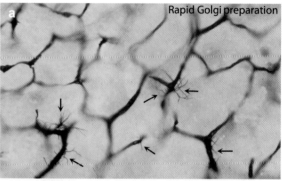

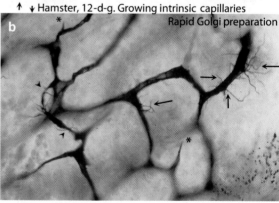

Fig. 7.9 Composite figure with two photomicrographs (**a**, **b**), from rapid Golgi preparations of the cerebral cortex of 12-day-old hamster embryos, showing the sprouting (growing) capillaries of its developing intrinsic microvascular system. Sprouting capillaries (*arrows*) are characterized by long filopodia emerging from the leading endothelial cells that project in all direction into the surrounding nervous system (**a**, **b**) and by the formation of connecting loops between contiguous capillaries (**b**, *arrowheads*). Projecting filopodia can also originate from endothelial cells of the capillary wall (**a**, **b**). Reabsorbing capillaries are characterized by single and very fine terminal (**b***)

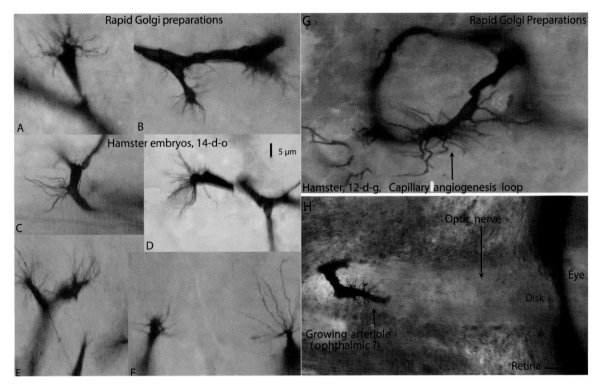

Fig. 7.10 Montage of several photomicrographs showing examples of sprouting (growing) capillaries, from rapid Golgi preparations of the cerebral cortex of 12- and 14-day-old hamster embryos. Newly formed and still growing capillaries are characterized by leading endothelial cells with projecting filopodia than fan-out into the surrounding tissue (**a–f**). The tip of these growing capillaries is characterized by a polyp-like enlargement formed by its leading endothelial cells (**a–f**). This type of growth is most commonly observed in the embryonic cerebral cortex. Growing capillaries (**c**) can also emerge directly from the vessel wall endothelial cells (see Fig. 7.9). Also illustrated (**h**) are the terminal endothelial filopodia of a vessel growing into the optic nerve of the eye of a 12-day-old hamster embryo. Also illustrated (**h**) are part of the retina and the optic nerve entrance disk. This growing vessel could represent a precursor of the ophthalmic artery

visible against the transparent (yellow-reddish) background of these preparations (Figs. 7.9a, b and 7.10a–f). The preparations offer nearly 3-D view of the sprouting capillaries and of its advancing (searching) filopodiae (Figs. 7.9a, b). Projecting filopodia occurs from the leading endothelial cells at the tip of growing capillaries (most frequent) as well as from endothelial cells of the capillary wall preceding the emergence of a new capillary (Fig. 7.9a, b and 7.10a–f). The radiation of filopodia of growing capillary suggests that they are searching the surrounding tissue for clues that will determine the directional growth of the emerging vessel. Rapid Golgi preparations also show the presence of reabsorbing capillaries (Fig. 7.9b *). A single, fine, short, and ending process characterizes the structure of reabsorbing capillary (Fig. 7.9b *). The occasional presence of capillary loops connecting contiguous capillaries at different depths has also been observed in some Golgi preparations (Figs. 7.9b, arrow heads). The activity (angiogenesis, reabsorption, and looping) of growing capillaries, illustrated in these Golgi preparations does not represent an isolated phenomenon but one observed throughout the cerebral cortex embryonic gray matter. Possibly, these Golgi illustrations represent the first time that this extraordinary vascular phenomenon has been demonstrated. Rapid Golgi preparations that include the embryo's entire head have permitted to observer the intrinsic angiogenesis of capillaries in other regions. Capillary angiogenesis seems to be essentially similar in all tissues studied. One extraordinary example of active capillary angiogenesis (filopodia formation) has been observed in a growing vessel advancing through the optic nerve of the eye of a hamster embryo (Fig. 7.10h). This early growing vessel could well represent the precursor of the future ophthalmic artery.

In the brain, the filopodia length of growing capillaries could range between 5 and 20 µm. The most spectacular ones are those formed at the tip of growing capillaries (Fig. 7.10a–f). The growing tip of a capillary forms a polyp-like structure composed of leading endothelial cells from where the filopodia radiate (Fig. 7.10a–f). Electromicroscopic observations of growing capillaries confirm this view (Fig. 7.4a, c). The leading endothelial cells slide over the parent vessel wall and afterward they become progressively incorporated into the new vessel wall (Fig. 7.4a, c). Capillary angiogenesis, reabsorption, and remodeling will continue to occur during the brain prenatal and postnatal maturation in response to functional demands.

How growing capillaries interconnect with each other and how they establish new interconnecting lumina between them are complex and poorly understood process. In experimental angiogenesis, the formation of new capillaries and the establishments of anastomoses are considered to be analogous processes (Cliff 1963). Both processes are characterized by the formation of accessory intercellular spaces among the leading endothelial cells filopodia and lamellopodia and by the secretion of a protein-rich material that feels them. These early intercellular spaces enlarge, coalesce, and eventually establish communication with the lumen of the parent vessel. Similar observations have been reported in tumor angiogenesis (Ausprunk and Folkman 1977; Ausprunk 1979; Folkman 1982), in inflammatory processes (Schoefl 1963; Ran and Majno 1977; Cotran 1982; Sholley et al. 1984) in corneal implants (Ausprunk 1979), in transparent chambers (Clark and Clark 1939), and in healing muscles wounds (McKinney and Panner 1972).

Rapid Golgi and electromicroscopic studies of the embryonic cortex intrinsic vascularization corroborate these observations (Marín-Padilla 1985, 1987, 1988). The endothelial filopodia of approaching growing capillaries are the first element to establish contacts. The leading filopodiae from approaching growing capillaries slide over each other establishing long narrow spaces filled with basal lamina material. Some of the contacting filopodia are progressively transformed into lamellopodia with enlargement of these narrow intercellular spaces. The original intercellular spaces between contacting lamellopodia enlarge by accumulating an increasing amount of basal lamina material and coalesce with each other, forming larger ones. The enlarging spaces formed between advancing endothelial cells of approaching capillaries are believed to participate in the formation of the intercommunicating lumen of the new capillary. However, how approaching capillaries establish the intercommunication between their respective lumina remains poorly understood. The luminal intercommunication between approaching growing capillaries is a rather complex process, impossible to be recognized and/or reconstructed using either rapid Golgi and/or electron-microscopic observations, which need to be further explored.

References

Andres KH (1976a) Zur Feinstruktur der Arachnoidalzotten bei Mammalia. Zeitschrift Zellsforschung Mikroskoscopic Anatomy 82:92–109

Andres KH (1976b) Über die Feinstruktur der Arachnoidea und Dura mater von Mammalia. Zeitschrift Zellsforschung Mikroskoscopic Anatomy 82:272–295

Ausprunk DH (1979) Tumor Angiogenesis. In: Houck JC (ed) Chemical messenger of the inflammatory process. North-Holland, Amsterdam, pp 317–351

Ausprunk DH, Folkman J (1977) Migration and proliferation of endothelial cells in preformed and newly formed blood vessels during angiogenesis. Microvasc Res 14:53–65

Bär T, Wolff JR (1972) The formation of capillary basal membranes and internal vascularization of the rat's cerebral cortex. Zeitschrift Zellsforschung Mikroskoscopic Zeitschrift Anatomy 133:231–248

Casley-Smith E, Földi-Börcsök E, Földi M (1971) The prelymphatic pathway of the brain as revealed by cervical lymphatic obstruction and the passage of particles. Br J Exp Pathol 57:179–188

Clark ER, Clark EL (1939) Microscopic observations of the growth of capillaries in the living mammal. Am J Anat 4:251–299

Cliff WJ (1963) Observations on healing tissue: a combined light and electron microscopic investigation. Philos Trans R Soc London Ser B 256:305–325

Cotran RS (1982) The endothelium and inflammation: new insights. In: Magno G, Cotran RS, Kaufman N (eds) Current topics in inflammation and infection. Williams and Wilkins, Baltimore, pp 18–37

Duvernoy HM, Delon S, Vannson JL (1981) Cortical blood vessels of the human brain. Brain Res Bull 7:519–579

Folkman J (1982) Angiogenesis: initiation and control. Annals New York Academy of Science 401:212–227

Hamilton WJ, Boyd JD, Mossman HW (1972) Human embryology. Heffer, Cambridge, UK

Jones EG (1970) On the mode of entry of blood vessels into the cerebral cortex. J Anat 106:507–520

Kaplan HA, Ford DH (1976) The brain vascular system. Elsevier, Amsterdam

Krahn V (1982) The pial mater at the site of the entry of blood vessels into the central nervous system. Anat Embryol 164:257–263

References

Krisch B, Buchheim W (1984) Access and distribution of exogenous substances in the intercellular clefts of the rat adenohypophysis. Cell Tissue Res 236:439–452

Krisch BW, Leonhardt H, Oksche A (1982) The meningeal compartment of the median eminence and the cortex: A comparative analysis in the rat. Cell Tissue Res 228:597–640

Krisch BW, Leonhardt H, Oksche A (1983) Compartments and perivascular arrangement of the meninges covering the cerebral cortex of the rat. Cell Tissue Res 238:459–474

Mall FP (1904) On the development of blood vessels of the brain in the human embryo. Am J Anat 4:1–18

Manasek FJ (1971) The Ultrastructure of embryonic myocardial blood vessels. Dev Biol 26:42–54

Marín-Padilla M (1985) Early vascularization of the embryonic cerebral cortex: Golgi and electron microscopic studies. J Comp Neurol 241:237–249

Marín-Padilla M (1987) Embryogenesis of the early vascularization of the central nervous system. In: Yasargil MG (ed) Microneurosurgery, clinical considerations and microsurgery of arteriovenous racemous angiomas. Stuttgart, Thieme-Verlarg, pp 23–47

Marín-Padilla M (1988) Embryonic vascularization of the mammalian cerebral cortex. In: Peters A, Jones AG (eds) cerebral cortex, vol VII. Plenum Press, New York, pp 479–509

Marín-Padilla M (1997) Developmental neuropathology and impact of perinatal brain damage II: white matter lesions of the neocortex. J Neuropathol Exp Neurol 56:219–235

Marín-Padilla M (1999) Developmental neuropathology and impact of perinatal brain damage III: gray matter lesions of the neocortex. J Neuropathol Exp Neurol 58:407–429

Marín-Padilla M, Parisi JE, Armstrong DL, Sargent SK, Kaplan JA (2002) Shaken infant syndrome: developmental neuropathology, progressive cortical dysplasia and epilepsy. Acta Neuropathol 103:321–332

McKinney RV, Panner BJ (1972) Regenerating capillary basement membrane in skeletal muscle wounds. Lab Investig 26:100–113

Morse DE, Low FN (1972) The fine structure of the pial mater of the rat. Am J Anat 133:349–368

Nabeshina S, Reese RS, Landis MD (1975) Junction of meninges and marginal glia. J Comp Neurol 164:127–170

O'Rahilly R, Muller F (1986) The meninges in human development. J Neuropathol Exp Neurol 45:588–609

Oda Y, Nakanishi I (1984) Ultrastructure of the mouse leptomeninx. J Comp Neurol 225:448–457

Padget DH (1948) The development of the cranial arteries in the human embryo. Contrib Embryol Carnegie Inst 32:207–261

Padget DH (1957) The development of the cranial venous system in man: from the point of view of comparative anatomy. Contrib Embryol Carnegie Inst 34:79–140

Pape EK, Wigglesworth JS (1979) In: Pape EK, Wigglesworth JS (eds) Hemorrhages, ischaemia and the perinatal brain. Spastics International Medical Publications, London, pp 11–38

Pessacq TP, Reissenweber NJ (1972) Structural aspects of vasculogenesis in the central nervous system. Acta Anat 81:439–447

Pile-Spellman JM, Mckusic KA, Strauss HW, Strauss Coony J, Taveras JM (1984) Experimental in vivo imaging of the cranial perineural lymphatic pathway. Am J Neuroradiol 5:539–545

Ran GB, Majno G (1977) Acute inflammation: a review. Am J Pathol 86:183–210

Sabin FR (1920a) Origin and development of the primitive vessels in the chick and the pig. Contrib Embryol Carnegie Inst 6:61–124

Sabin FR (1920b) Study of the origin of blood vessels and of red blood corpuscles as seen in the living blastoderm of chick during the second day of incubation. Contrib Embryol Carnegie Inst 9:213–262

Schoefl GI (1963) Studies on inflammation III. Growing capillaries: their structure and permeability. Virchows Arch 337:97–141

Sholley MM, Ferguson GP, Seibel HR, Montour JL, Wilson JD (1984) Mechanism of neovascularization, vascular sprouting can occur without proliferation of endothelial cells. Lab Investig 51:625–642

Streeter GL (1918) The developmental alterations in the vascular system of the brain in the human embryo. Contrib Embryol Carnegie Inst 8:5–38

Strong LH (1964) The early embryonic pattern in internal vascularization of the mammalian cerebral cortex. J Comp Neurol 123:121–138

8 Human Motor Cortex First Lamina and Gray Matter Special Astrocytes: Development and Cytoarchitecture

During the development of the mammalian cerebral cortex, the gliogenesis of the different glial cell types parallel the evolution of its various compartments. Each basic glial cell type evolves at specific time, occupies a specific compartment, and develops specific morphological and functional features. The following glial cell types are recognized in the developing mammalian cerebral cortex: radial glia cells, white matter fibrous astrocytes and oligodendrocytes, subependymal polymorphous astrocytes, first lamina special astrocytes, and the gray matter protoplasmic astrocytes. The developmental history of each type is characteristic and occurs at specific time. The developmental histories and morphologic features of most glial cells types have been well documented (Rakic 1972, 1988; Marín-Padilla 1995). However, the late developmental gliogenesis of both the first lamina astrocytes and the gray matter protoplasmic astrocytes are less well documented. First lamina special astrocytes are needed for the late maintenance of the cerebral cortex expanding external glial limiting membrane (EGLM) and the protoplasmic astrocytes are necessary for the late gray matter neuronal and vascular maturations (Marín-Padilla 1995).

The late developmental histories of both the first lamina special astrocytes and the gray matter protoplasmic astrocytes are closely interrelated and are independent from that of other cortical glia cell types. Their gliogenesis takes place late in development and in response to specific requirements of the expanding first lamina and the maturation of the gray matter. The late developmental expansion of the cortex EGLM requires of a continued incorporation of additional glial endfeet, which are no longer provided by the preexisting radial glial cells. Moreover, the late developmental regression of radial glial cells will make the gliogenesis of first lamina special astrocytes imperative, both during late prenatal and postnatal cortical maturations. The gray matter late ascending neuronal and vascular maturations require the gliogenesis of special protoplasmic astrocytes.

The late gliogenesis of these two new glial cell types are interlinked and related to the subpial granular layer (SGL) of Ranke (Ranke 1909; Brun 1965; Marín-Padilla 1995). In the human cerebral cortex, the late developmental appearance and subsequent disappearance of the SGL coincide with the expansion of the first lamina, the ascending maturation of the gray matter neuronal and intrinsic microvascular systems, and the late gliogenesis of these two glial cell types. These developmental histories are interrelated and dependent on each other.

The current idea concerning the mammalian cerebral cortex gliogenesis proposes that all glial cell types originated from radial glial cell transformations (Rakic 1972, 1984, 1988; Schmechel and Rakic 1979; Levitt et al. 1983; Pixley and De Bellis 1984; Fedoroff 1986; Kadhim et al. 1988; Voigt 1989; Choi 1990; Culican et al. 1990; Cameron and Rakic 1991). However, the late gliogenesis of some glial cell types is a much complex process concomitant with the establishment of different compartments and is, essentially, independent of the original radial glial cell system. From a developmental perspective, the human cerebral cortex gliogenesis could be separated into an early, an intermediate, and a late phase. The early gliogenesis in the cerebral cortex is represented by the generalized establishment of the radial glial cell system; the intermediate phase by the gliogenesis of white matter fibrous astrocytes and of oligodendrocytes; and, the late phase by the gliogenesis of first lamina special astrocytes and gray matter protoplasmic astrocytes. The glial elements of each cortical compartment evolve at specific times and all originate independently from the cortex ependymal epithelium (Marín-Padilla 1995).

The developing human cerebral cortex early gliogenesis is characterized by the formation of the radial glial cell system. These cells evolve from the ventricular cells of the original neuroectoderm and are characterized by several terminal filaments with endfeet that united by tight-junction construct and maintain the cortex EGLM and manufacture its basal lamina (Chaps. 2–7). These cell's long filaments extend, uninterruptedly, from ependymal to pial surfaces and eventually play a crucial role in the ascending transportation of neuronal precursors up to the first lamina (Rakic 1984; Marín-Padilla 1992). The radial glial cell filaments lengthen progressively, without losing either their ependymal and/or pial attachments, paralleling the developing cortex increasing thickness. From the 7th to around the 15th week of gestation, the main functional role of the radial glial is the transportation of pyramidal neuron precursors from the ependymal epithelium up to the first lamina. These precursors will participate in the ascending and stratified formation of the gray matter pyramidal cell plate (PCP) (Chaps. 3 and 4). Rapid Golgi preparations, demonstrate the radial glial cells uninterrupted filaments and the progressive incorporation of their terminal endfeet into the cortex expanding EGLM (see Fig. 3.9 of Chap. 3 or Fig. 7.5a of Chap. 7). In addition to the early developmental transportation of pyramidal neuron precursors for the maturing gray matter, the radial glial cells will also participate, in my opinion, in another important and late transportation of glial cell precursors up to the first lamina (Marín-Padilla 1995). This late transportation of glial precursor up to first lamina, via radial glial filaments, participates in the formation of the ephemeral subpial granular layer (SGL) of Ranke. Eventually, the SGL cells become the source of both first lamina special astrocytes and of gray matter protoplasmic astrocytes (Marín-Padilla 1995).

The developing human cerebral cortex intermediate gliogenesis is characterized by the incorporation of fibrous astrocytes and of oligodendrocytes into the expanding white matter compartment. The gray matter, at this time, is completing its PCP formation and still is functionally immature and avascular (Chaps. 3 and 7). During this time, the ependymal epithelium, in addition to radial glial cells, have numerous other cell types with complex ascending processes of various lengths (see Fig. 3.9b of Chap. 3 or Fig. 7.5a of Chap. 7). Eventually, these elements lose their ependymal attachment and become free-ascending glial cell precursors destined to the expanding white matter. They are quite abundant throughout the thickness of both paraventricular and matrix zones. They ascend, free and independent from the radial glial filaments, up to the white matter zone. As they reach and penetrate into the white matter zone, their leading processes bend in opposite directions, accompanying either corticipetal and/or corticofugal fibers. These free-ascending glial cell precursors, of ependymal origin, become the source of both white matter fibrous astrocyte and oligodendrocytes (Marín-Padilla, 1995). The white matter fibrous astrocytes are characterized by long radiating filaments, without bifurcations, which establish vascular contacts through specialized endfeet. Shorter processes and direct association with the white matter fibers characterize the oligodendrocytes. During intermediate developmental stages, the number of free-ascending glial precursors is considerable throughout the paraventricular and expanding white matter zones. Moreover, this germinal zone is considered to be the source of both neuronal and glial precursors (Levitt et al. 1983). The incorporation of free-ascending glia cell precursors into the white matter also coincides with the development of its microvascular system. The migration of free-ascending glial precursors into the white matter compartment will cease when mitotic activity of preexisting fibrous astrocytes and/or oligodendrocytes will generate the needed additional elements.

In rapid Golgi preparations, the subependymal paraventricular zone (germinal layer) is characterized by the presence of large polymorphous astrocytes that are associated with its large, irregular, and thin-walled anastomotic vessels (see Fig. 7.5a of Chap. 7). This type of large polymorphous astrocytes is not found anywhere else in the developing human cerebral cortex (Marín-Padilla, 1995). During the matrix zone active period (between the 20th and the 30th week of gestation) its large anastomotic vessels are particularly vulnerable to hypoxia resulting in ruptures and local hemorrhages with devastating clinical results (Marín-Padilla, 1996). By the 30th week of gestation only remnants of the paraventricular germinal zone are recognized in the human cerebral cortex and by birth time this zone in no longer recognizable. The late prenatal involution of the paraventricular germinal matrix zone involves the regression of both its vascular anastomotic plexus and its polymorphous astrocytes.

The developing human cerebral cortex late-gliogenesis coincides with the ascending and stratified maturation of the gray matter PCP and is characterized by the establishment of both first lamina astrocytes and gray matter protoplasmic astrocytes. To understand the developmental need for a late-gliogenesis, various features of the developing human cerebral cortex should be considered. During the early developmental stages, radial glial fibers provide the necessary endfeet that, joined by tight junction, build and maintain the neocortex EGLM and manufacture its basal lamina. However, the EGLM late expansion will require the continued incorporation of additional glial endfeet, which are no longer provided by the preexisting radial glia fibers. On the other hand, the increasing distance between the developing gray matter and the ependymal epithelium caused by late expansion of the white matte will make the migration of free-ascending glial precursors increasingly difficult, time consuming, and eventually impossible. Consequently, the need for a new and faster route for glia cell precursors to reach the developing cortex outer strata (first lamina and gray matter) becomes imperative. The ascending migration of neuronal precursors – via radial glia fibers – destined to the formation of the gray matter pyramidal cell plate starts around the 7th week of gestation and is nearly completed by 15th week. Subsequently, the gray matter PCP starts the ascending and stratified maturations of its neuronal and intrinsic microvascular systems (see Chaps. 3 and 7). Both developmental processes require an uninterrupted incorporation of new glial elements.

In the human motor cortex, from the 15th to around the 30th week of gestation, the existing radial glial fibers participate, in my opinion, in a massive ascending migration of glial cell precursors destined to the first lamina. This late ascending migration of glial cell precursors (via radial glia fibers) establishes, within the first lamina, the subpial granular layer (SGL) of Ranke (Ranke 1909; Brun 1965; Marín-Padilla 1995). The SGL small undifferentiated cells start to accumulate within the first lamina, around the 15th week of gestation, reach their maximal accumulation by the 25th week, afterward their number start to diminish and by the 30th week of gestation, they are verily recognizable. The cerebral cortex matrix (germinative) zone, after the completion of the neuronal migration, produces most of the glial cell precursors (Gressens et al. 1992). In my opinion the SGL cells are the source of first lamina special astrocytes and, subsequently, some of these astrocytes become the source of gray matter protoplasmic astrocytes. The transformation of first lamina astrocytes into gray matter protoplasmic astrocytes is a remarkable event recently described using the rapid Golgi procedure (Marín-Padilla 1995).

The enigmatic SGL appearance and subsequent disappearance, its composition, and its developmental role have remained the source of controversy. According to Ranke (1909) original description, its cells migrate into the neocortex gray matter and become glial cells; according to Brun (1965), some cells undergo dilution, some disappear, others migrate inwardly to become either glia cells or neurons, while still others may participate in some pathological conditions (leptomeningeal heterotopias); according to Gadisseux et al. (1992), its cells become first lamina special neurons; and, according to Meyer and Gonzales-Fernandez (1993), some SGL cells become neurons, others glial cells, while many others degenerate after the 24th week of gestation. In my opinion, the SPL represents a subpial accumulation of glial cell precursors that reach the first lamina late in development, via radial glial fibers, and become the source of first lamina special astrocytes and, subsequently, of gray matter protoplasmic astrocytes, which concurs with Ranke's original proposition (Marín-Padilla 1995). How the SGL cells evolve into first lamina astrocytes and, subsequently, into gray matter protoplasmic astrocytes are explored below using the rapid Golgi procedure.

The SGL developmental history is characterized by three successive events: (a) progressive subpial accumulation of undifferentiated glial cells precursors, which ascend from ependymal to pial surface reusing, as guides, preexisting radial glia filaments; (b) successive transformation of these undifferentiated cells into first lamina astrocytes, which provide the additional endfeet needed for the maintenance of the expanding EGLM and for the manufacture of its basal lamina; and, (c) progressive inward migration of first lamina astrocytes into the developing gray matter and their subsequent transformation into protoplasmic astrocytes. Notably, the timing of these processes coincides with the ascending and stratified neuronal and microvascular maturations of the gray matter PCP (see Chaps. 3 and 7). The inward migration of first lamina astrocytes and subsequent transformation into protoplasmic astrocytes start after the 16th week of gestation and continues, uninterruptedly, through the neocortex late prenatal development (Marín-Padilla 1995). The eventual mitotic activity of preexisting first lamina astrocytes and gray matter protoplasmic astrocytes will supply the additional elements needed during the cerebral cortex postnatal maturation.

8.1 First Lamina Special Astrocytes

The first lamina unique special astrocytes (comet cells) have been known since classical times. Retzius (1894) and Ramón y Cajal (1891) provided excellent morphologic descriptions of them using the Golgi procedure. In vertically cut rapid Golgi preparations, first lamina astrocytes appear as small cells with short horizontal, curved, and ascending processes that terminate in endfeet incorporated into the cortex EGLM (Fig. 8.1b, c). Their soma and processes are covered by short and fine excrescences and/or lamellae, which resemble those of the gray matter protoplasmic astrocytes (Fig. 8.1c, d). They tend to occupy the first lamina upper zone (Fig. 8.1a–c). In tangentially cut rapid Golgi preparations, they appear as small multipolar cells with horizontal branches, parallel to the pial surface, covered by short and fine lamellae (Fig. 8.2d). They appear to intermingle with numerous axonic fiber terminals coursing in all directions, probably representing afferent fibers (see Chap. 4). During the neocortex lateprenatal development, these astrocytes are quite numerous throughout the entire cerebral cortex subpial zone. The large number of first lamina astrocytes and the incorporation of their endfeet into the EGLM can be also demonstrated in glial fibrillary acidic protein (GFAP) preparations (Fig. 8.1e).

Some first lamina astrocytes, while still attached to the EGLM by endfeet, have descending branches that penetrate into the gray matter and establish vascular contacts with developing capillaries (Fig. 8.1a, *arrows*). During late prenatal development, the number of first lamina astrocytes with descending branches that penetrate into the maturing gray matter increases progressively, becoming a universal feature throughout the cortex (Figs. 8.1–8.5). First lamina astrocytes with descending processes into the gray matter become the source of protoplasmic astrocytes.

Many first lamina astrocytes, throughout the cortex, while still attached to the EGLM by endfeet grow long descending processes (Figs. 8.1a, 8.2a, b, 8.3a–d, 8.4a–c). The length of some descending processes reaches up to 350 µm (Fig. 8.4c). It is important to emphasize, that from the 15th week of gestation to the time of birth, the distance between the first lamina and the last maturing pyramidal cell strata and intrinsic microvascular system ranges roughly between 300 to 400 µm (Chaps. 3 and 7). This distance remains essentially unchanged during the prenatal maturation of the human cerebral cortex. It roughly represents the thickness of the still undifferentiated and avascular gray matter PCP upper strata (Chap. 3). The descending processes of first lamina astrocytes must travel this distance to reach the capillaries of the intrinsic microvascular system capillaries that accompany the pyramidal neurons ascending maturation (Chap. 3).

8.2 Grey Matter Protoplasmic Astrocytes

The progressive transformation of first lamina astrocytes into gray matter protoplasmic astrocytes is a remarkable developmental event, which occurs, uninterruptedly, from the 16th to the 35th week of gestation

Fig. 8.1 Montage of photomicrographs (**a–d**), from rapid Golgi preparations of the motor cortex of 30-week-old human fetuses and (**e**) a glial fibrillary acidic protein stain preparation (EGLM) from an adult brain, showing the morphological features of first lamina special astrocytes. (**a–c**) Reproduce first lamina vertically cut Golgi preparations showing the morphologic features of its special astrocytes with ascending processes with endfeet incorporated into the external glial limiting membrane (EGLM) and, some of them, with descending processes that have penetrated into the underlying gray matter. They are covered by short and fine excrescences and lamellae, which resemble those of gray matter protoplasmic astrocytes. The descending processes of some first lamina astrocytes have established vascular contacts (*arrows*) with gray matter capillaries. The astrocytes are mainly located throughout the first lamina upper region and above the Cajal–Retzius axon terminals (C-R at). (**d**) Illustrates a first lamina tangentially cut Golgi preparation showing several special astrocytes (G) with multiple processes running parallel to pial surface that are intermingled with crisscrossing axonic fiber terminals. (**e**) First lamina vertically cut GFAP preparation showing numerous small astrocytes scattered throughout its upper zone with terminal endfeet incorporated into the neocortex EGLM

8.2 Grey Matter Protoplasmic Astrocytes

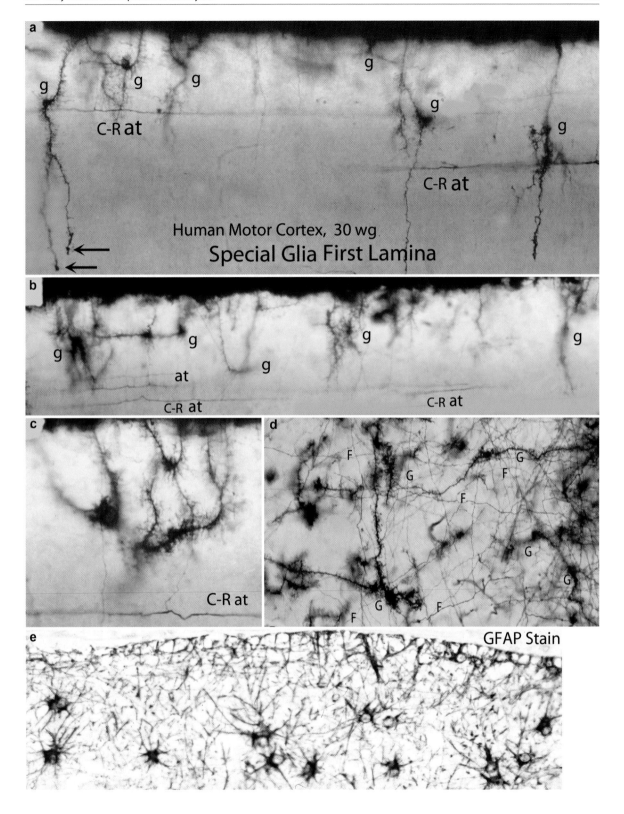

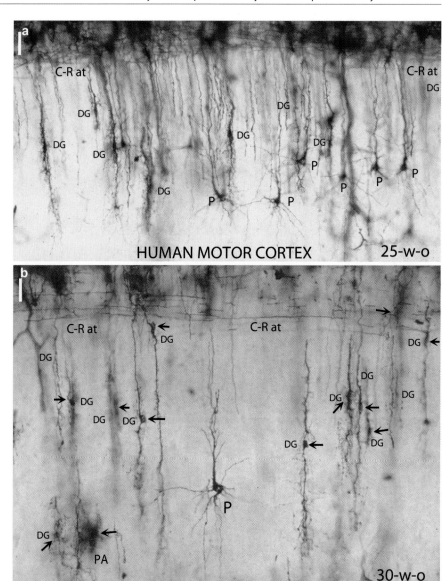

Fig. 8.2 Montage with two photomicrographs, from rapid Golgi preparations of the motor cortex of 25- (**a**) and 30-week-old (**b**) human fetuses, showing numerous descending glial processes (DG) from first lamina astrocytes intermingled with the apical dendrites of pyramidal neurons (P). Some descending glial processes are from first lamina astrocytes with endfeet still attached to the EGLM; others (DG) already detached from the EGLM are free-descending glial elements, while some of the deepest ones (PA) are already transformed into gray matter protoplasmic astrocytes. The nuclei of these transforming glial elements could occupy different levels (*arrows*). (**b**) Also illustrates, at a higher magnification, several first lamina astrocytes attached to the EGLM as well as some descending ones and Cajal–Retzius cells (C-R at) long horizontal axon

(Marín-Padilla 1995). This development event is clearly demonstrated in rapid Golgi preparations of the human motor cortex late gestational stages (Figs. 8.2–8.5). Three distinct stages are recognized in the sequential transformation of first lamina astrocytes into gray matter protoplasmic ones (Figs. 8.2–8.6).

First stage (1): Some first lamina astrocytes, while still attached to the EGLM by endfeet, develop descending processes that enter into the gray matter (Figs. 8.2a, b, 8.3a–d, 8.4a–c). These early protoplasmic astrocytes precursors have long trailing processes that continue to be attached to the EGLM by endfeet (Figs. 8.2b, 8.3b, d, 8.4a–c, 8.5a, c). These processes are covered by fine and short lamellae (Figs. 8.2a, b, 8.3a, c). Their nucleus can be found at any level within the descending process (Figs. 8.2b, 8.3a–d, 8.4a, c, 8.5a–d, *short arrows*). The nucleus of some descending astrocyte may still be located within first lamina (Fig. 8.4c, *arrow*). The descending process of early protoplasmic astrocytes, while still attached to the EGLM, could measure up to 350 µm in length (Fig. 8.4c).

Second stage (2): Eventually, descending astrocytes lose their EGLM attachment, retract their trailing process, and become free protoplasmic astrocytes

8.2 Grey Matter Protoplasmic Astrocytes

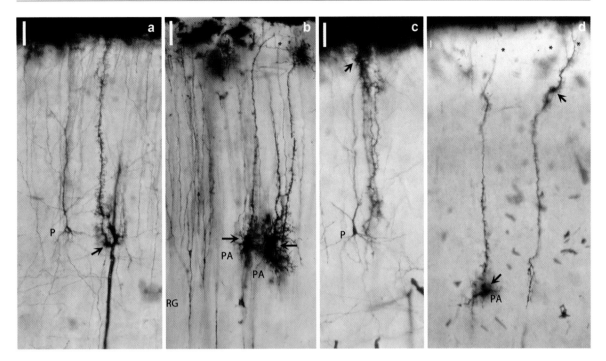

Fig. 8.3 Montage of several photomicrographs, from rapid Golgi preparations of the motor cortex of 28- (B) and 30-week-old (a, c, d) human fetuses, showing, at a higher magnification, several mature protoplasmic astrocytes (PA) among pyramidal neurons (P), some with long trailing processes (a, b, d), and other with a trailing process still attached (*) to the cortex EGLM. Also illustrated are first lamina (I) astrocytes (c, d) with long descending processes with the nuclei (arrows) still within the first lamina. B: Also illustrates several radial glial fiber terminals (RG) with long ascending branches that terminate in endfeet incorporated into the cortex EGLM

(Figs. 8.2b, 8.4b, 8.5a). Their nucleus also descends within their trailing process (Figs. 8.3b, 8.5a). Free protoplasmic astrocytes start to develop radiating processes covered by short lamellae and establish vascular endfeet contacts with local capillaries (Figs. 8.2b, 8.3c, d, 8.4a, b, 8.5a–c). Many intermediate stages between first and second stages are recognized throughout the maturing cerebral cortex.

Third stage (3): Eventually, the gray matter protoplasmic astrocyte become stellate cells, develop radiating, often bifurcated, processes covered by lamellae that terminate forming endfeet contacts with the newly developing capillaries of the gray matter (Figs. 8.3b, 8.4a, b, 8.5a–d). The number of vascular contacts of any mature protoplasmic astrocyte increases progressively (Fig. 8.5d). Within the same Golgi preparation, it is not infrequent to observe simultaneously descending protoplasmic astrocytes precursors still attached to the EGLM by endfeet, descending ones with disconnected trailing processes and mature ones (Figs. 8.2b, 8.3b, 8.5a, c). The association of mature protoplasmic astrocytes with the capillaries of the intrinsic microvascular system increases progressively (Fig. 8.5c).

The gray matter protoplasmic astrocytes' triple developmental stages are recognized from the 16th week of gestation to the time of birth and are particularly prominent from the 18th to the 30th week of gestation. It is important to point out that the differentiation of protoplasmic astrocytes is an ascending and stratified process that parallels the gray matter neuronal and microvascular ascending maturations. The transformation of first lamina astrocytes into gray matter protoplasmic astrocytes will continue until mitotic activity of preexisting ones will generate the additional ones needed for the human cerebral cortex postnatal maturation. The deepest and older protoplasmic astrocytes of the gray matter are the first to undergo mitotic activity.

Camera lucida reconstructions, from rapid Golgi preparations, of 20-, 25- and 35-weeks-old human fetuses, recapitulate, to scale, the sequential transformation of

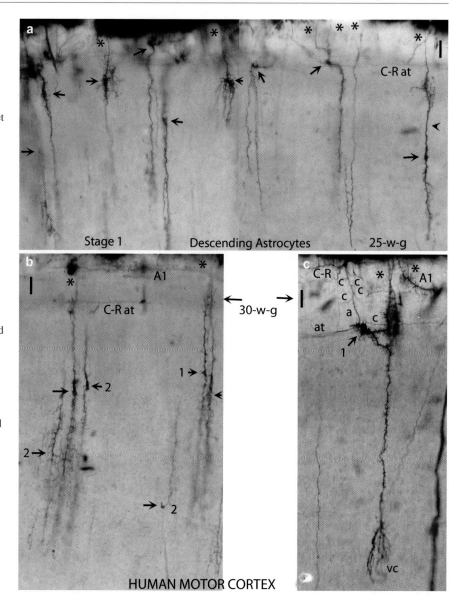

Fig. 8.4 Montage of photomicrographs, from rapid Golgi preparations of the motor cortex of 25- (**a**) and 30-week-old (**b, c**) human fetuses showing the morphologic features of first lamina astrocytes still endfeet attached to the EGLM with long descending processes that have penetrated into the gray matter zone. Some first lamina astrocytes (**a**) are at stage 1 (1) of their transformation and still attached to the EGLM while others (**b**) are at developmental stage 2 (2) and already disconnected from it. C: Illustrates, at a higher magnification, a first lamina astrocyte (stage 1) still attached (*) to the EGLM, with its nucleus (arrow) within the lamina and a long descending process that has already established vascular contacts (vc) with gray matter capillaries. Also illustrated is a Cajal–Retzius neuron (C-R) with a descending axon with several collaterals (c) and its horizontal axonic (at) terminal

first lamina astrocytes into gray matter protoplasmic astrocytes (Fig. 8.6). Their progressive triple transformation is illustrated at each developmental age (Fig. 8.2 (1), (2,) (3)) Also, the distance travelled by the descending maturing astrocytes remains essentially unchanged. This distance reflects that which existed, at any age, between the first lamina and the last formed intrinsic capillary plexus and pyramidal cell maturation within the gray matter (Marín-Padilla 1985). This distance also coincides with the length of the last maturing pyramidal cells apical dendrites and with the thickness of the still undifferentiated PCP (Figs. 8.2a, b, 8.3c). See also Chaps. 3 and 7. The extraordinary transformations of these glia cell precursors is not an isolated event but one that occurs throughout the entire cerebral cortex from the 16th week of gestation to around the time of birth (Fig. 8.2a, b).

The first lamina astrocyte's participation in the maintenance of the expanding cortical EGLM increases

8.2 Grey Matter Protoplasmic Astrocytes

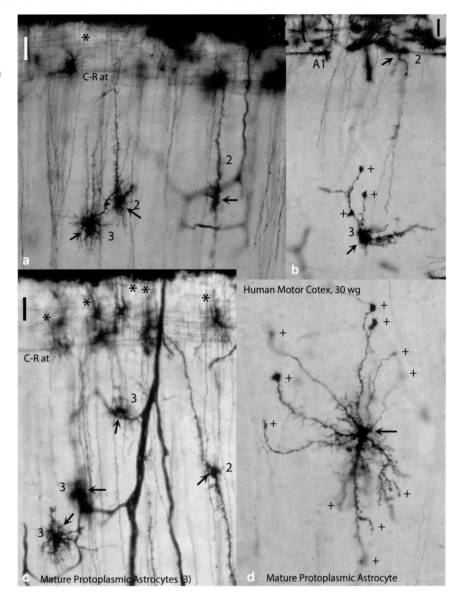

Fig. 8.5 Montage of four photomicrographs (**a**, **b**, **c**, **d**), from rapid Golgi preparation of the motor cortex of 30-week-old human fetuses, showing various stages (stage 1 in A, stage 2 in **a** & **c**, and stage 3 in **b** & **d**) of the transformation of first lamina astrocytes into gray matter protoplasmic astrocytes. The newly developed gray matter protoplasmic astrocytes are closely associated with the gray matter intrinsic microvasculature (**c**) and some of them (**b**, **d**) have already established vascular contacts (+). The first lamina fibrillar organization and Cajal–Retzius cells axon terminals (C-R at) are also illustrated in **a** and **c**

progressively as that of the radial glial decreases late in development. Late in prenatal cortical development, the number of radial glial fibers decreases progressively and by birth time only a few of them are recognizable. After birth, first lamina astrocytes are the only elements that contribute endfeet to the brain expanding EGLM. Since there is no additional ascending migration of glial cell precursors, the mitotic activity of preexisting first lamina astrocytes will maintain an adequate glial cell population for the EGLM postnatal maintenance and reparation. It should also be emphasized that cortex gray matter gliogenesis and vasculogenesis are concomitant, ascending, and stratified events, which parallel the ascending and stratified maturation of its pyramidal neurons.

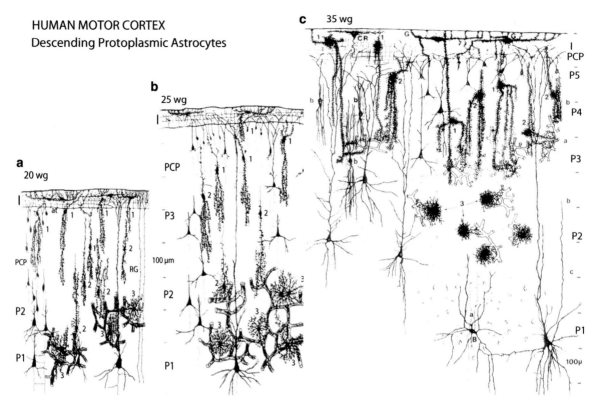

Fig. 8.6 Composite figure with three camera lucida drawings, from rapid Golgi preparations of the motor cortex of 20- (**a**), 25- (**b**) and 35-week-old (**c**) human fetuses, recapitulating comparatively at similar scale, the three (1, 2, 3) developmental stages of the transformation of first lamina astrocytes into gray matter protoplasmic astrocytes. The developmental transformation of first lamina astrocytes into gray matter protoplasmic astrocytes remain essentially identical throughout the human cerebral cortex late gestational ages. The descending distance traveled by these astrocytes to reach the ascending gray matter microvascular system also remain unchanged though cortical development. This distance also represents the thickness of the still undifferentiated pyramidal cell plate (PCP), which these descending glial cells must cross to reach the gray matter vessels. Also illustrated is the gray matter ascending stratification and maturation of its various pyramidal cells strata, of its intrinsic microvascular system association to the maturing protoplasmic astrocytes. The illustration's additional features are self-explanatory. The 100-μm scales are roughly applicable to all drawing

References

Brun A (1965) The subpial granular layer of the foetal cerebral cortex in man. Its ontogenesis and significance in congenital cortical malformations. Acta Pathologica Microbiologica Scandinavica 13(Suppl 179):1–98

Cameron RS, Rakic P (1991) Glial cell lineage in the cerebral cortex: a review and synthesis. Glia 4:124–137

Choi BH (1990) Gliogenesis in the developing human fetal brain. In: Levi G (ed) Neurology and neurobiology, Differentiation and function of glial cells, vol. 55. Wiley, New York, pp 6–16

Culican SN, Baumrind M, Yamamoto H, Pearlman A (1990) Cortical radial glial: identification in tissue cultures and evidence for their transformation to astrocytes. J Neurosci 10:648–692

Fedoroff S (1986) Prenatal ontogenesis of astrocytes. In: Fedoroff S, Vernadakis A (eds) Astrocytes, morphology, and regional specialization of astrocytes, vol. 1. Academic, New York, pp 35–74

Gadisseux J-F, Goffinet AM, Lyon G, Evrard P (1992) The human transient subpial granular layer: an optical, immunohistochemical, and ultrastructural analysis. J Comp Neurol 324:94–114

Gressens PC, Richelme HJ, Gadisseux J-F, Evrard P (1992) The germinative zone produces the most cortical astrocytes after neuronal migration in the developing mammalian brain. Biol Neonate 61:4–24

Kadhim HJ, Gadisseux J-F, Evrard P (1988) Topographical and cytological evolution of the glia during prenatal development of the human brain: a histochemical and electron microscopic study. J Neuropathol Exp Neurol 47:166–188

References

Levitt P, Cooper ML, Rakic P (1983) Early divergence and changing proportions of neuronal and glial precursor cells in the primate cerebral ventricular zone. Dev Biol 96:472–484

Marín-Padilla M (1992) Ontogenesis of the pyramidal cell of the mammalian neocortex and developmental cytoarchitectonic: a unifying theory. J Comp Neurol 321:223–240

Marín-Padilla M (1985) Early vascularization of the embryonic cerebral cortex. Golgi and electron microscopic studies. J Comp Neurol 241:237–249

Marín-Padilla M (1995) Prenatal development of fibrous (White matter), protoplasmic (gray matter) and layer I astrocytes in the human cerebral cortex: a golgi study. J Comp Neurol 357:554–572

Marín-Padilla M (1996) Developmental neuropathology and impact of perinatal brain damage. I. Hemorrhagic lesions of the neocortex. J Neuropath Exp Neurol 55:746–762

Meyer G, Gonzales-Fernandez T (1993) Developmental changes in layer I of the human neocortex during prenatal life: a DiI-tracing and AChE and NADPH-d histochemistry study. J Comp Neurol 338:317–336

Pixley S, De Bellis J (1984) Transition between immature radial glia and mature astrocytes studied with a monoclonal antibody to vimentin. Dev Brain Res 15:201–209

Rakic P (1972) Mode of cell migration to the superficial layers of the fetal monkey neocortex. J Comp Neurol 145:61–84

Rakic P (1984) Emergence of neuronal and glia cell lineages in primate brain. In: Black IB (ed) Cellular and molecular biology of neuronal development. Plenum, New York, pp 29–50

Rakic P (1988) Intrinsic and extrinsic determinants of neocortical parcellation: a radial unit model. In: Rakic P, Singer W (eds) Neurobiology of the neocortex. Wiley, New York, pp 92–114

Ramón y Cajal S (1891) Sur la structure de l'ecorce cerébrale de quelques mammiferes. La Céllule 7:125–176

Ranke O (1909) Beiträge zur Kenntnis der normalen und pathologischen Hirnrindenbildung. Beitrage für Pathology Anatomy 45:51–85

Retzius G (1894) Die Neuroglia des Gehirns beim Menschen und bei Såugetieren. Biologie Unters 6:1–109

Schmechel DE, Rakic P (1979) A Golgi study of radial glial cells in developing monkey telencephalon; morphogenesis and transformation into astrocytes. Anat Embryol 156:115–152

Voigt T (1989) Development of glial cells in the cerebral wall of ferrets: direct tracing of their transformation from radial glia into astrocytes. J Comp Neurol 289:74–88

New Developmental Cytoarchitectonic Theory and Nomenclature

What might distinguish the human brain from that of other mammals has been, since classical times, in the mind of neuroscientists and philosophers. Although different opinions have been advanced, none has survived scientific scrutiny. Despite obvious brain differences among mammalian species, the problem remains unsolved. From a developmental perspective, the present monograph explains, for the first time, why, how, and in what order the mammalian cerebral cortex becomes stratified and, also, how many strata distinguish each species (Fig. 9.1). Only a comparative study of the developing mammalian cerebral cortex, with an appropriate method, will be able to determine the nature and evolution of its distinguishing laminar stratification. The present chapter describes developmental rapid Golgi observations concerning the cytoarchitectural organization and ascending sequential stratification of the developing motor cortex, essentially, of humans and also of cats, hamsters, and mice. Moreover, the incorporation of corticipetal fibers, of non-pyramidal neuronal systems, of inhibitory neurons, of the formation of the intrinsic microvascular system, and the maturation of protoplasmic astrocytes are all concomitant ascending processes that parallel the ascending stratification of the various pyramidal cell functional strata of the gray matter, where most neurons reside (Chaps. 2–8). The establishment of an additional P6 pyramidal cell stratum distinguishes the human motor cortex as human and, therefore, different from that of other primates. Consequently, the idea, universally accepted, that the only cytoarchitectural difference in the cerebral cortex among mammalian species is one of degree but not of kind need to be re-evaluated.

In addition, the mammalian cerebral cortex current theory of descending laminations (Layers I, II, III, IV, V, VI, and VII), found in most embryology, neuroanatomy, and neurohistology textbooks, also need to be re-evaluated. This classic conception evolves from the studies of Broadmann, His, Cajal, Vignal, Koelliker, Retzius, Braak, and many others. Some of the laminar variations are understood as modifications of a particular lamination (Broadmann 1909; Lorente de Nó 1922, 1949; Poliakov 1961, 1964; Blinkov and Glezer 1968; Shkol'nik-Yarros 1971; Kemper and Galaburda 1984). However, the current concept of descending laminations has serious difficulties in explaining satisfactorily some of these variations (Marín-Padilla, 1978, 1992, 1998). The present chapter also challenges the current theory of descending laminations and its nomenclature as well as the idea that apical dendrites of pyramidal neurons, from any lamination, ascend up to the first lamina (Chaps. 3 and 4).

The current theory of descending cortical laminations confronts insurmountable problems in explaining the numerical differences in the layering found in the cerebral cortex of some mammals (Marín-Padilla 1978). The solutions offered to explain these differences are without scientific and/or developmental support. A few examples from the literature should illustrate these particulars (Fig. 9.2). In order to retain the classic seven descending laminations, Poliakov (1964), in describing the European hedgehog cerebral cortex, condenses into a single II-III-IV stratum, three (layers II, III, and IV) laminations (Fig. 9.1). Similarly, Lorente de Nó (1922), in the mouse cerebral cortex, condenses the second (layer II) and third (layer III) laminations of classic nomenclature, into a single II-III stratum (Fig. 9.1). Moreover, to resolve the increment in the number of laminations found in higher mammalian species, the third (layer III) lamina has been either duplicated (IIIa and IIIb) in the monkey and/or triplicate (IIIa, IIIb and IIIc) in the human cerebral cortex (Lorente de Nó, 1949; Bailey and von Bonin 1951;

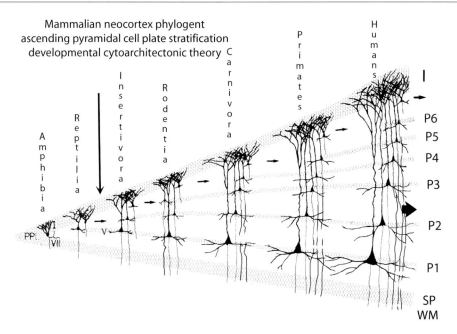

Fig. 9.1 Schematic representation of the new developmental cytoarchitectonics theory and nomenclature applicable to all mammalian species, introduced in the present chapter. From and evolutionary perspective, the theory proposes that the cortex gray matter pyramidal cell plate (PCP) is a mammalian innovation as well as a biologically open nucleus capable of incorporating additional pyramidal cell strata to those already formed for controlling the increasing motor capabilities of the evolving mammals. The PCP is originally formed and expands within a primordial cortical organization (PP), shared with amphibian and reptiles, from which the cortex first lamina (I) and the subplate (SP) zone evolve. The figure illustrates schematically the proposed increase in the number of pyramidal cell strata that could occur during mammalian evolution. From an evolutionary perspective, the theory proposes that the postnatal maturation of the residual PCP, which characterizes the motor cortex of all mammals, including humans (and possibly reptiles?), may participate in the establishment of the additional pyramidal cell stratum that characterizes each species (*arrows*). In humans, the postnatal maturation of the residual PCP could participate in the establishment of an additional P7 pyramidal cell stratum in the adult cortex. *PP*, primordial plexiform; *I*, first lamina; *SP*, subplate zone; P1, P2, P3, P4, P5, and P6 represent the various ascending pyramidal cell strata that characterize the evolving cerebral cortex of the different mammalian species; *WM*, white matter. (Modified from Marín-Padilla 1998)

Poliakov, 1961; Blinkov and Glezer 1968; Shkol'nik-Yarros 1971; Jones 1975). Arbitrary nomenclatures have been also used to designate the human cerebral cortex's various laminations (Fig. 9.2). It appears that seven distinct laminations only distinguish the cat's (carnivores) cerebral cortex, as this chapter demonstrates (Chap. 10). The lack of understanding of why the mammalian cerebral cortex is stratified (laminated) might explain the confusion and the use of arbitrary nomenclatures. Also, tradition, the larger-than-normal names of some of the supporting neuroscientists, and the absence of developmental studies are also contributing factors to the present situation. As I can gather, the current (classic) theory of descending cortical laminations remains unsupported by developmental data. To correctly interpret why, how, and in what order the mammalian cerebral cortex is stratified (laminated) its developmental history must be first understood.

A complete developmental study of the mammalian cerebral cortex as well as the use of an appropriated method capable of staining all its neuronal, fibrillary, synaptic, microvascular, and glial systems will be required to settle this problem. The rapid Golgi procedure and the use of unspoiled postmortem brain tissue, is undoubtedly the best procedure, possible the only, capable of staining all components of the developing cerebral cortex and appreciate their interrelationships (Chapter 12).

The present chapter introduces the results, of lifelong rapid Golgi studies of the prenatal development of the motor cortex and humans, cats, hamsters, and mice embryos and fetuses. The fundamental observations made, include: (a) the cortex gray matter (where most neurons reside) stratification is a progressive ascending process from lower (older) to upper (younger) strata (Chaps. 3 and 4); (b) the incorporation of afferent fiber terminals into the developing gray matter also follows an ascending progression that parallel the ascending maturation (stratification) of its pyramidal neurons (Chap. 3); (c) the incorporation and functional

Fig. 9.2 Table showing several examples of the variable and arbitrary nomenclatures used to describe the variable number of cortical laminations (strata) assigned to various mammalian species. The main features of the table are self-explanatory. (From Marín-Padilla 1978)

Hedgehog	Mouse	Cat	Monkey	Man Child	Adult	Adult	Adult	Adult	Adult
1	I	I	I	I	I	I	I	I	1
II + III + IV	II + III	II	II	II	II	II	II	II	II
V	IV	III	IIIa	IIIa	III	IIIa	IIIa	IIIa	IIIa
VI	V	IV	IIIb	IIIb	IVa	IIIb	IIIb	IIIb	IIIb
VII	VIa	V	IV	IIIc	IVb	IIIc	IIIc	IIIc	IIIc
	VIb	VI	V	VI	Va	IV	IV	IV	IV
		VII	VI	V	Vb	V	Va	Va	Va
			VII	VIa	Vc	VIa	Vb	Vb	Vb
				VIb	VIa	VIb	VI	VI	VIa
					VIb	VII	VII	VII	VIb
Poliakov, 1964	Lorente de Nó, 1922	Marin-Padilla, 1977*	Jones, 1975	Shkol'nic-Yarros, 1971	Lorente de Nó, 1949	Poliakov, 1961	Bailey, von Bonin, 1951	Blinkov and Glezer, 1968	Blinkov and Glezer, 1968

maturation of the various excitatory-inhibitory neuronal systems also follows an ascending progression that parallel the ascending pyramidal neurons strata formation (Chap. 6); (d) the development of the cortex intrinsic microvascular systems (Chap. 7) and the differentiation of the gray matter protoplasmic astrocytes (Chap. 8) are interrelated and concomitant processes that also follow an ascending progression from lower to upper pyramidal cell strata; and, (e) the Cajal–Retzius cells of first lamina play a significant role in the ascending placement of the various pyramidal cells strata (Chap. 5). Consequently, the development of all essential components of the mammalian cerebral cortex follow an ascending and stratified maturation, from deeper (older) to upper (younger) strata, which contrast sharply with the current conception of descending functional laminations and with its nomenclature.

The original working hypothesis for these developmental Golgi studies was based on the following assumptions: first, the motor activity of each mammalian species should be reflected in the cytoarchitectural complexity and stratification of its motor cortex; and second, the mouse motor cortex cytoarchitectural organization should be simpler than that of a cat and that of the cat should be simpler than that of a human. Such that the mouse cerebral cortex cytoarchitectural organization should be distinguished as mouse, that of a cat as cat, and that of a human as human. Because of these propositions, the rapid Golgi studies were dedicated almost exclusively to explore the developmental history of the cerebral cortex motor region of these mammals. The developmental rapid Golgi studies presented in this chapter corroborate both assumptions. The chapter introduces a new developmental cytoarchitectonics theory and a new nomenclature, applicable to the cerebral cortex of all mammalian species, and challenges the current theory of descending laminations and its nomenclature (Fig. 9.1).

The new cytoarchitectonics theory proposes: first, that mammalian cerebral cortex cytoarchitectural and functional organizations are essentially and primarily based on its pyramidal neurons; second, that the number of pyramidal cell strata established in the motor cortex reflects the motor proficiency of each species and therefore varies among them; and third, that the incorporation and functional maturation of non-pyramidal neurons, of afferent fibers and of the intrinsic microvascular and neuroglial associated systems are concomitant and parallel to the pyramidal cell ascending maturation and stratification (Fig. 9.1). It also proposes that the pyramidal neuron is a mammalian innovation and the cortex essential functional system (Chaps. 3 and 4). Consequently, the mammalian cerebral cortex cytoarchitectural organization should be envisioned as a series of ascending – from older (deeper) to younger (upper) – pyramidal cell functional strata rather than as a series of descending laminations as the current theory proposes (Fig. 9.1). The new nomenclature describes the cerebral cortex cytoarchitecture as a series of ascending pyramidal cell functional strata – from older (deeper) to younger (upper) ones, which are herein renamed P1, P2, P3, P4, etc., reflecting the order of their ascending maturation as well as the number of strata finally established (Fig. 9.1). The correspondence between the current and the new nomenclatures is comparatively illustrated in some of the figures (Fig. 3.14 of Chap. 3 and Fig. 4.2 of Chap. 4).

The new cytoarchitectonics theory envisions the mammalian cerebral cortex pyramidal cell plate (PCP) as its most distinguishing feature as well as a mammalian innovation (Chap. 4). The mammalian cerebral cortex gray matter is considered to be a single, stratified, and biologically open nucleus capable of adding new pyramidal cells strata as new motor activities evolve in the course of mammalian phylogeny (Fig. 9.1). Also, that the mammalian gray matter PCP is formed within a primordial cortical organization, which is shared with amphibian and reptiles, representing the cortex first lamina and the subplate (SP)zone (Fig. 9.1). The appearance of the PCP establishes, simultaneously, the cerebral cortex first lamina, above it, and the subplate zone, below it, each with its characteristic neuronal and fibers compositions (Chaps. 2 and 3). The first lamina and SP zone neurons share structural and functional interrelationships and together they constitute the mammalian neocortex primordial functional system. This primordial functional system will be replaced when the pyramidal neurons, in conjunction with the Cajal–Retzius cells, assume the control of cerebral cortex principal and definitive functional system. The first lamina basic neuron, the Cajal–Retzius cell, will persist throughout cortical development and will continue to play a fundamental role in its function (Chap. 4). On the other hand the subplate neurons, after the 15th–16th weeks of gestation, start to lose their connections with first lamina and to regress both morphological and functionally. Eventually, they become transformed into deep interstitial neurons (Chap. 3). Around the 15th–17th week of gestation, the human cerebral cortex starts to assume it definitive function based of its pyramidal neurons (Chaps. 3 and 4).

The theory proposes that the ascending maturation of each pyramidal cell stratum is escorted by functional contacts with ascending afferent fiber terminals (Chap. 3), the incorporation of excitatory-inhibitory neuronal systems (Chap. 6), the ascending and stratified maturation of the intrinsic microvascular system (Chap. 7), and by the ascending differentiation of gray matter protoplasmic astrocytes (Chap. 8). Also, that the mammalian cerebral cortex is essentially composed of two basic types of neurons, namely: those that are developmental, morphological, and functionally constrained by their attachment to first lamina (pyramidal neurons) and those that are free from such constrains (local-circuit interneurons). The first group is represented by numerous, fixed, stable, similar among mammals, and excitatory pyramidal neurons of various sizes, which are primarily involved in the management of the animal essential motor needs (e.g., locomotion in search of food and mate and in avoiding danger). The second group is represented by the less numerous, less stable, unfixed, morphologically variable, and functionally changeable local-circuit interneurons (mostly inhibitory cells). Local-circuit interneurons are capable of reacting, changing, and adapting morphologically and functionally to intrinsic and/or extrinsic influences and trespass these modifications into the functional activity of pyramidal neurons. While pyramidal neuron's morphologic appearance remain unchanged viewed from any perspective, the morphological appearances of local-circuit interneurons, which are spatially (3D) oriented within the cortex, varies depending of the angle of view (Marín-Padilla and Stibitz 1974; Marín-Padilla 1990a, b).

The central nervous system (in general) and the cerebral cortex (in particular) are systems that evolve exclusively for controlling the animal motor activities (Llinás 2003). Their evolution is characterized by the progressive formation of ascending (caudal-cephalad) motor nuclei (strata) destined to control the animal increasing motor activities and capabilities. The last CNS nucleus established is the mammalian cerebral cortex. The mammalian cerebral cortex is also a sequentially stratified organization, from older to younger pyramidal cells strata. The new theory further proposes that the number of pyramidal cell strata formed in the mammalian cerebral cortex is species-specific and that, in general, reflects the motor capabilities of each one (Fig. 9.1). In general terms, it appears, from histological studies, that the number of pyramidal cell strata seems to increase from the European hedgehog to humans, perhaps reflecting the increasing complexity of their motor activities (Ramón y Cajal 1911; Lorente de Nó 1922, 1949; Bailey and von Bonin 1951; Sholl 1956; Poliakov 1961, 1964; Romer 1966; Blinkov and Glezer 1968; Ebner 1969; Marín-Padilla 1970a, b, 1978, 1992, 1998; Shkol'nik-Yarros 1971; Valverde 1983; Morgane et al. 1990).

Excluding the first lamina and the deep polymorphous neurons (layer VII in current nomenclature), the actual number of distinct pyramidal cell strata observed in the hedgehog cortex gray matter is two, three in the mouse, four in the cat, five in primates, and six in humans

(Fig. 9.1). It would seem that the hedgehog, for controlling all its motor activities, only needs two pyramidal cell strata, the mouse three, the cat four, the primates five, and humans, at least, six strata (Figs. 9.1 and 9.2).

The new theory further proposes that the last pyramidal cell strata incorporated into the cortex control the inherited as well as the newly acquired (learned) motor activities that each species acquired during postnatal life. In this context, the motor cortex, in all mammals that I have been able to explore, including humans, have an additional thin stratum, immediately beneath first lamina, composed of numerous small and undifferentiated pyramidal neurons (Chaps. 3 and 4). This stratum represents a remnant of the original PCP and its functional maturation will occurs postnatally. From an evolutionary perspective, this uncommitted neuronal stratum could contribute, postnatally, to the acquisition of new pyramidal cell strata added in the course of mammalian evolution (Fig. 9.1, *small arrows*).

The new theory further supports two additional features pertaining to the pyramidal neurons functional attributes. The first one pertains to the sequential and ascending functional maturation of pyramidal neurons (still functionally anchored to first lamina) as they sequentially establish contacts with ascending corticipetal fibers (Chap. 4). The second one pertains to the pyramidal neuron eventual descending functional activity (Chap. 4). During prenatal development, pyramidal neurons grow by sequentially adding functionally active membrane to their apical dendrite, for the formation of synaptic contacts, without losing their anchorage to first lamina (Chap. 4). At 15 weeks of gestation, a pyramidal neuron of stratum P1 (the first established in the human motor cortex) measures 270 μm in length, has short basal dendrites, and may have one or two dendritic spines (see Fig. 4.1 of Chap. 4). By birth time, A P1 pyramidal neuron length is around 1,500 μm, has many long basal and collaterals dendrites, and thousands of spines cover its apical dendrite alone (Chap. 4). At 27 day of gestation, a P1 pyramidal neuron of the cat motor cortex measures 25 μm and may have one or two dendritic spines; by birth time, its length is about 670 μm and may have up to 700 spines in its apical dendrite (Chap. 11). At 13 day of gestation, the length of a P1 pyramidal neuron of the hamster motor cortex is about 15 μm and by birth its length has increased up to 600 μm and its apical dendrite may have up to 420 spines (Marín-Padilla 1992). The P1 pyramidal neurons eventual functional activity depends on a descending functional pathway that interconnects sequentially upper to lower pyramidal neurons (Ramón y Cajal 1933; Weiler et al. 2008), such that a P1 pyramidal neuron receives descending inputs from all pyramidal neurons of the above strata and its functional output will reflect the number of participating pyramidal cell strata (Chap. 3). The local-circuit interneurons also participate in this descending functional pathway. The more strata are incorporated into in the mammal motor cortex the more complex the functional output of its P1 pyramidal neurons.

The new cytoarchitectonic theory submits that the human motor cortex is characterized by the presence of an additional P6 pyramidal cell stratum (Fig. 9.3). This stratum distinguishes the human motor cortex as

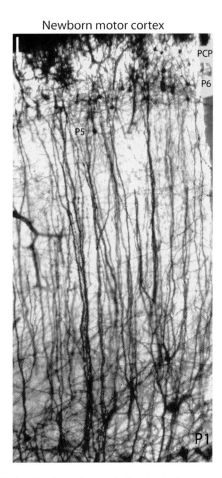

Fig. 9.3 A color photomicrograph showing the large number apical dendrites from deep pyramidal neurons (stratum P1) that reach and branch within the first lamina (I) and the numerous small pyramidal neurons of stratum P6 (P6) from a rapid Golgi preparation of the motor cortex of a newborn infant. Also illustrated are some of the small and still undifferentiated neurons of the residual pyramidal cell plate (PCP). The presence of a P6 pyramidal cell stratum distinguishes the human motor cortex as human and different from that of other primates. Some pyramidal neurons of stratum P1 and P5 are also illustrated. Compare this figure with Figs. 4.1 and 4.3 of Chap. 4, which also illustrate the overall cytoarchitectural organization of the newborn motor cortex

human and hence different from that of any other primates. It is located below the first lamina, extends throughout the entire cortical surface, and is composed of innumerable closely packed small pyramidal neurons, ranging in length from 40 to 50 μm (Fig. 9.4). They are all characterized with short apical dendrites that branch profusely within first lamina and by rich axonic profiles (Fig. 9.4). Their extensive dendritic and axonic terminals establish a complex mesh of neuronal interconnections that extends throughout the human brain's entire surface (Figs. 9.3 and 9.4). Moreover, there are more neurons is this stratum than there are in the rest of the cerebral cortex (Fig. 9.4). The pyramidal neurons of stratum P6 are also accompanied by small basket, Martinotti, double tufted, and chandelier inhibitory interneurons. The pyramidal cell P6 stratum large number of neurons, their cytoarchitectural and functional complexity and universal expansion throughout the cerebrum are distinct features of the human cerebral cortex (see also

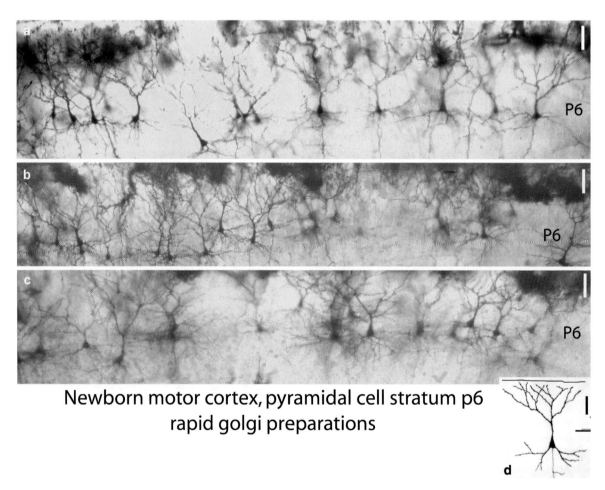

Fig. 9.4 Montage of photomicrographs (**a–c**), from rapid Golgi preparations of the motor cortex of newborn infants, showing, at a higher magnification, the morphological features and the large number of small pyramidal neurons of the P6 pyramidal cell stratum. These neurons are characterized by short apical dendrites with rich terminal dendritic bouquets and by their distinct stratification. In the newborn motor cortex, the length (size) of P6 pyramidal neurons ranges between 40 and 50 μm, compared to the 1,500-μm lengths of those of stratum P1. Their dendrites are covered by dendritic spines and possibly by other types of synaptic contacts. Their dendritic and axonic territories form a dense and complex interconnecting functional mesh that extends throughout the entire cerebral cortex. There are more pyramidal neurons in the P6 stratum than there are in the rest of the cerebral cortex. The numerous pyramidal cells of P6 stratum and their complex and extensive functional territory could represent the anatomical substratum for the unique motor and creative capabilities that distinguish the human species. The inset (**d**) reproduces a camera lucida drawing illustrating, in detail, the P6 pyramidal neurons basic morphological features and location immediately below the first lamina

Fig. 3.14a, b of Chap. 3 and Figure 4.2a of Chap. 4). The extension of the P6 pyramidal cells throughout the entire brain surface makes this stratum the most likely anatomical substratum for the control of the uniquely motor capabilities that distinguish the human species.

In recent genetic studies, I have read that in some DNA segments the base-sequence is uniquely human (Pollard et al. 2006). Some of these segments seem to be involved in the development of the human brain (HAR1), on its large number of neurons and large size (ASPM), on human speech (FOXP2), and, on the development of the hand-opposing thumb (HAR2), which plays an important role in the human motor capabilities and creativity. Although unfamiliar with these genetic studies, they seem to give particular support to some of the developmental observations presented in this chapter. Per example, the human DNA accelerated regions (HAR1) seems to be expressed in both C-R cells and in reelin, from around the 7th to the 19th week of gestation, which coincide with the ascending and stratified formation of the motor cortex pyramidal cell plate as described in this chapter.

Finally, the new cytoarchitectonic theory further proposed that the postnatal maturation of the residual PCP could incorporate an additional P7 pyramidal cell stratum into the human motor cortex (Fig. 9.1, *large arrow*). This additional pyramidal cell P7 stratum could, in conjunction with P6, represent the anatomical substratum for the higher motor and mental capabilities (thoughts, inventiveness, motor creativity, wisdom, intelligence, and consciousness) that distinguish the human species.

References

Bailey P, von Bonin G (1951) The isocortex of man. University of Illinois Press, Urbana

Blinkov SM, Glezer II (1968) The human brain in figures and tables. Plenum, New York

Broadmann B (1909) Verleichende Lokalisationslehre der Grosshirnrinde in iheren Prizipien Dargestellt auf Grund des Zellenbaues. Barth, Leipzig

Ramón y Cajal S (1911) Histologie du Systéme Nerveux de L'Homme et des Vertébrés. Maloine, Paris

Ramón y Cajal S (1933) Neuronismo o Reticularismo? Las pruebas objetivas de la unidad anatómica de las células nerviosas. Madrid (Reprinted by the Cajal Institute in 1952)

Ebner FF (1969) A comparison of primitive forebrain organization in metatherian and eutherian mammals. Ann NY Acad Sci 167:241–257

Jones EG (1975) Lamination of differential distribution of thalamic afferents within the sensory motor of the monkey. J Comp Neurol 160:167–204

Kemper TLB, Galaburda AM (1984) Principles of cytoarchitectonic. In: Peters A, Jones EG (eds) Cerebral cortex, Cellular components of cerebral cortex, vol. 1. Plenum, New York, pp 35–54

Llinás R (2003) El Cerebro y el Mito del Yo. Barcelona, Belacqua Ediciones

Lorente de Nó R (1922) Corteza cerebral del ratón. I. La corteza acústica. Trabajos del Laboratorio de Investigaciones Biológicas 20:41–78

Lorente de Nó R (1949) Cerebral cortex. Architecture, intracortical connections, motor projection. In: Fulton JF (ed) Physiology of the nervous system. Oxford University Press, Oxford, pp 274–313

Marín-Padilla M (1970a) Prenatal and early postnatal ontogenesis of the human motor cortex. A Golgi study. I. The sequential development of the cortical layers. Brain Res 23:167–183

Marín-Padilla M (1970b) Prenatal and early postnatal ontogenesis of the human motor cortex. A Golgi study. II. The basket-pyramidal system. Brain Res 23:185–191

Marín-Padilla M (1978) Dual origin of the mammalian neocortex and evolution of the cortical plate. Anat Embryol 152:109–126

Marín-Padilla M (1990a) Three-dimensional structural organization of layer I of the human cerebral cortex: a Golgi study. J Comp Neurol 229:89–105

Marín-Padilla M (1990b) The pyramidal cell and its local-circuit interneurons: A hypothetical unit of the mammalian cerebral cortex. J Cogn Neurosci 2:180–194

Marín-Padilla M (1992) Ontogenesis of the pyramidal cell of the mammalian neocortex and developmental cytoarchitectonics: a unifying theory. J Comp Neurol 321:223–240

Marín-Padilla M (1998) Cajal-Retzius cell and the development of the neocortex. Trends Neurosci 21:64–71

Marín-Padilla M, Stibitz G (1974) Three-dimensional reconstruction of the baskets of the human motor cortex. Brain Res 70:511–514

Morgane PJ, Glezer II, Jacobs MS (1990) Comparative and evolutionary anatomy of the visual cortex of the dolphin. In: Jones EG, Peters A (eds) Cerebral cortex, Comparative Structure and Evolution of cerebral Cortex, vol. 8B. Plenum, New York, pp 215–262

Poliakov GI (1961) Some results into the development of the neuronal structure of the cortical ends of the analyzers in man. J Comp Neurol 117:197–212

Poliakov GI (1964) Development and complications of the cortical part of coupling mechanism in the evolution of vertebrates. J Für Hirnforshung 7:253–273

Pollard KS, Salama SR, Lamber N, Lambot M-A, Coppens S, Pedensen JS, Katzmen S, King B, Onodera C, Siepez A, Kern AD, Dehay C, Igel H, Ares M, Vanderhaeghen P, Hausler D (2006) An RNA gene expressed during cortical development evolves rapidly in human. Nature 443:167–172

Romer AS (1966) Vertebrate paleontology. Saunders, Philadelphia

Shkol'nik-Yarros E (1971) Neurons and interneuronal connections of the central nervous system (English translation by Haigh B). Plenum, New York, pp 35–58

Sholl DA (1956) The organization of the cerebral cortex. Methuen, London

Valverde F (1983) A comparative approach to neocortical organization based on the study of the brain hedgehog (*Erinaceous europaeus*). In: Grisolia S, Guerri C, Samson F, Norton S, Reinoso-Suárez F (eds) Ramón y Cajal contributions to the neurosciences. Elsevier, Amsterdam, pp 149–170

Weiler N, Wood L, Yu J, Solla SA, Shepherd GMG (2008) Top-down laminar organization of the excitatory network in motor cortex. Nat Neurosci 11:360–366

Epilogue 10

The present monograph introduces new developmental data elucidating, for the first time, why, how, and in what order the mammalian cerebral cortex becomes stratified as well as the number and timing of the different strata formation (Chaps. 3–9). The ascending maturation of pyramidal neurons represents a mammalian innovation and the cerebral cortex essential functional axis. These neurons' anatomical and functional maturations follow, in all mammals, an ascending and stratified progression from lower (and older) to upper (and younger) gray matter regions. Moreover, the functional incorporation into the developing cortical gray matter of the different fiber systems, local-circuit interneurons, intrinsic microvascular system, and differentiation of protoplasmic astrocytes all follow an ascending and concomitant stratification that parallels with pyramidal neurons (Chaps. 3–8). The mammalian cerebral cortex cytoarchitectural organization should be thought as a series of ascending and functionally interconnected strata (laminations) rather than by descending ones as the current and universally held opinion proposes. A new nomenclature has also been introduced that reflects the cortex ascending anatomical and functional stratification, which should replace that of descending laminations (Layers I, II, III, IV, V, VI, and VII) currently used (Chaps. 3–9).

The developmental observations presented in the monograph support the following: (a) the mammalian cerebral cortex is an open biological system capable of adding new pyramidal cell functional strata, to those already established, for controlling the acquired motor activities that characterizes the evolving mammalian species; and, (b) the number of pyramidal cell strata established in the cerebral cortex reflects the motor ability and capabilities that characterize each mammalian species and vice versa. Consequently, the mouse motor cortex structural and functional complexity should be simpler and less stratified than that of the cat, and that of the cat should be simpler and less stratified than that of human. Cajal already anticipated this idea: "It is presumed that the extend and complexity of the gray matter (cerebral cortex) are closely related to the psychological hierarchy of each mammalian species" (Cajal 1911). Others have expressed similar ideas: "Prediction of motor activity, (*is*) the cerebral cortex primordial function" (Llinás, 2003); "There is, then, no genuine action if there is no thought, and there is no authentic thought if it is not duly referred to action." (Ortega and Gasset 1957); and, "Man paints with his cerebrum not with his hands" (attributed to Miguel Angel).

The presence of an additional P6 pyramidal cell stratum in the motor cortex of the newborn infant represents a fundamental observation (Chaps. 3–9). This P6 stratum distinguishes the human motor cortex as human and different from that of other primates. It could participate in the control of those unique motor capabilities that distinguish the human species such as thinking, speaking, writing, painting, sculpturing, poetizing, making and playing music, practicing sports, and other postnatally acquired motor activities. Another basic observation refers to the fact that the motor cortex of newborn human infants, as well as that of all mammalian species that I have been able to study, is also characterized by the presence of a thin band, immediately under the first lamina, of residual and still undifferentiated pyramidal cell plate (Chaps. 3–9). The functional maturation of the pyramidal neurons of this residual plate occurs during postnatal life, since it is no longer recognizable during late infancy. Its maturation could result in the postnatal incorporation of an additional P7 pyramidal cell stratum into the human motor cortex. It is known that the cerebral cortex of the newborn has one less stratum than the adult brain (Shkol'nik-Yarros 1971; Lorente de Nó 1949; Poliakov 1961). From an evolutionary perspective, the postnatal maturation of the residual PCP, which characterizes the motor cortex of all mammals (which I have studied), could participate in the establishment and incorporation of the additional pyramidal cell strata observed during mammalian evolution (Chap. 8).

I envision the human thoughts as traversing – back and for – through the innumerable functional interconnections established among the numerous small pyramidal neurons of stratum P6 and eventually of stratum P7, receiving different sensory inputs as they cross the brain's various regions and capable, at any time, of reactivating the motor cortex neurons for a specific motor action and/or motor creativity. In this extensive network setting, human thoughts remain active, independent, and ready for transformation into motor creativity, by acting through the motor cortex pyramidal system, when desired.

The functional interactivity of the pyramidal neurons of both P6 and eventually P7 strata, throughout the cerebral cortex, could represent the anatomical substratum for the human brain intrinsic electric activity. Early electroencephalographic studies demonstrated the presence of an intrinsic electric activity, even at rest, throughout the human brain (Berger 1929). This intrinsic activity has been recently reconfirmed by functional magnetic resonance imaging studies (Fox Raichle 2007, Raichle 2010). Human thoughts, traversing through the extensive interconnecting plexus of the cortex outer pyramidal cell P6 and P7 strata, could represent the anatomical substratum for this intrinsic activity and, speculating further, of human consciousness. Also the functional activity of P6 and P7 pyramidal cell strata could represent the anatomical substratum for the so-called Brain's Dark Matter and of human consciousness (Raichle 2010).

Considering the developmental observations presented in this monograph, the Darwinian idea of continuity of mind between humans and other animals, supported by some, may no longer be sustainable (Darwin 1936). The persona, who each one of us is, may reside in his/her brain pyramidal cell stratum P6 and, postnatally, in conjunction with stratum P7. The functional activity of these two pyramidal cell strata could represent the anatomical substratum for thoughts, intelligence, inventiveness, motor creativity, wisdom, compassion, heroism, altruism, contentment, and consciousness, which, among others, represent unique and distinctly human attributes.

The above-mentioned human attributes transcend the domain and/or sphere of biology and are immune to any of its laws. The capacity of human thoughts to add, subtract, establish new interconnections and decide novel motor activities involving painting, sculpturing, writing, poetizing, making and playing music and/or sports, as well as been human, are written in the history of humanity. Although strictly unnecessary for biological subsistence, the postnatal development of P6 and P7 pyramidal cells strata could provide humans with extraordinary intellectual, mental, and humanitarian capabilities. By developing and using these additional strata, *Homo sapiens* may be the only mammalian species capable, during his existence, of ascending (though briefly and rarely) into the status of *Homo humanus*. Paraphrasing Cajal's last words. While aphasic by a stroke, he wrote these words: "You have ... a supreme organ of knowledge and adaptation, ... And your inquisitive capabilities are far from been exhausted; rather, they may increase continuously, *such that every evolving new phase of Homo sapiens might exhibit the characteristic of a new humanity* "(the italics are mine). These "final words" of the Nobel Prize winning investigator, were published in a Madrid periodical, after his death (Fernandez Santarén 1997).

From scientific, philosophical, and cultural perspectives, others have expressed comparable thoughts: "Any human can be, if he want to, the result of its own brain" (Cajal 1911); "A human is the only reality that does not simply consist in being but must choose its own being" (Ortega and Gasset 1957) and Bronowski in *The Ascent of Man* (Bronowski 1974).

References

Berger H (1929) Über das Elektronenkephalogramm des Meschen. Archive fúr Psychiatrie Nervenkrankheiten 87: 527–570

Bronowski J (1974) The ascent of man. Little Brown, New York

Darwin, C (1936) The descent of man. The Modern Library, New York

Fernandez Santarén J (1997). El legado de Cajal a la Academia. Revista de la Real Academia de Ciencias, vol. 91. Madrid

References

Fox MD, Raichle ME (2007) Spontaneous fluctuations in brain activity observed with functional magnetic resonance imaging. Nat Rev Neurosci 8:700–711

Lorente de Nó R (1949) Cerebral cortex architecture, intracortical connections, motor projection. In: Fulton JF (ed) Physiology of the nervous system. Oxford University Press, Oxford, pp 274–313

Llinás R (2003) El Cerebro y el Mito del Yo. Belacqua Ediciones, Barcelona

Ortega Y, Gasset J (1957) Man and people. Norton & Company, New York

Poliakov GI (1961) Some results into the development of the neuronal structure of the cortical ends of the analyzers in man. J Comp Neurol 117:197–212

Raichle ME (2010) The brain's "dark energy". Sci Am 302: 44–49

Ramón Y Cajal S (1911) Histologie du Systéme Nerveux de l'homme et des Vertébrés, vol. 2. Maloine, Paris

Shckol'nic-Yarros EG (1971) Neurons and interneural connections of the central visual system (English Translation from Russian by Basil Haigh). Plenum Press, New York

Cat Motor Cortex: Development and Cytoarchitecture

The original developmental rapid Golgi studies of the cat cerebral cortex introduced a novel idea concerning the mammalian neocortex dual origin (Marín-Padilla 1971). The idea proposed that the mammalian neocortex evolves from a primordial cortical organization, shared with amphibian and reptiles, to which an expanding pyramidal cell plate (PCP) was added. The PCP is a mammalian innovation capable of adding additional pyramidal cell strata that reflect the animal motor capabilities. Several years later, this idea was corroborated, also in the developing cat cerebral cortex (Luskin and Shatz 1985) and, today, is universally accepted (Marín-Padilla 1978, 1998).

The cat Golgi developmental studies were possible by unique circumstances. Dr. D.E. Colby, Dartmouth Medical School Veterinarian, was interested in cat time-pregnancies for research purpose and requested my collaboration knowing my interest in embryonic development. He tries to induce estrus in young female cats by injecting mare's serum gonadotropin (Colby 1970). The injections were successful, mating occurs and the gathering of cat embryos of desired gestational age possible.

The cat cerebral cortex prenatal development, as well as that of other mammals, is characterized by an early embryonic and late fetal period. The cat embryonic period is characterized by four sequential stages: the undifferentiated neuroepithelium (19-d-o), the marginal zone (20-d-o), the primordial plexiform zone (20-d-o), and the appearance of the pyramidal cell plate (25-d-o). The developing cytoarchitecture of the cat neocortex during the embryonic period has been analyzed in this monograph second chapter. The early embryonic period is nearly identical in all mammals studied, including humans (See Chapters 2 and 3). The cat motor cortex fetal cytoarchitectural development, as it evolves through the 30th-, 35th-, 40th-, 45th-, 50th-, and 55th-day gestation and of newborns is explored in this chapter using rapid Golgi preparations. Cats are born around the 60th day of gestation. The quality and revealing capabilities of the selected rapid Golgi illustrations of the developing cat motor cortex speaks for themselves.

It is important to emphasize that the Golgi observations of the prenatal development of the cat motor cortex, presented in this chapter, corroborate those of the prenatal development of the human motor cortex (See Chapter 3).

11.1 Cat 25-Day-Old Stage

The cat neocortex fetal development starts, around the 25th day of gestation, with the appearance of the PCP, between the first lamina and the subplate zone. At this age, the newly established PCP is composed or two to three levels of pyramidal neurons with smooth apical dendrites anchored into the first lamina (see Fig. 2.3a, b). The cat neocortex is also characterized by a first lamina with Cajal–Retzius (C-R) cells and a large subplate SP zone with pyramidal-like neurons with dendrites and recurrent axon collaterals that reach first lamina. Martinotti cells with ascending axons that reach first lamina are also recognized.

11.2 Cat 30-Day-Old Stage

At this age, the cat neocortex mid region thickness is about 300 μm (Fig. 11.1a). It is composed of a well-developed first lamina with C-R cells (Fig. 11.1c), a three to four cells thick PCP (Fig. 11.1e), an SP zone

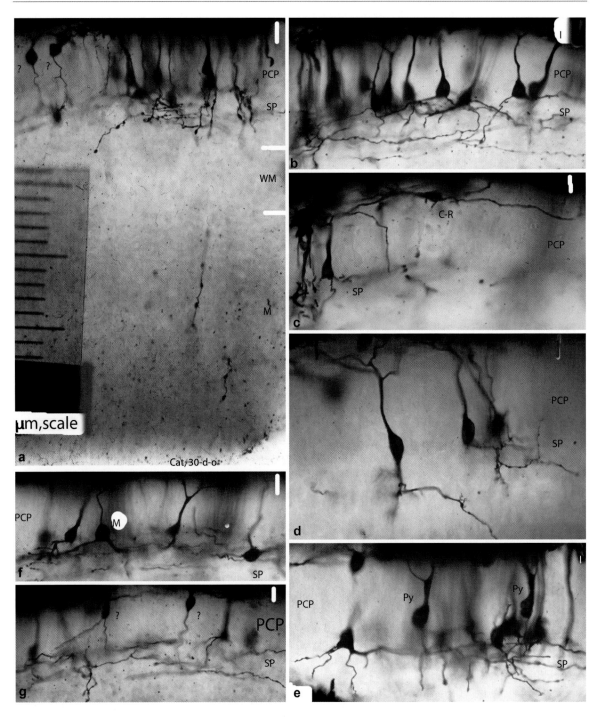

Fig. 11.1 Montage of photomicrographs showing several views of the mammalian cerebral cortex early cytoarchitectural organization and composition of 30-d-o cat fetuses from rapid Golgi preparations. Most of the neurons illustrated represent subplate (SP) zone pyramidal-like neurons with ascending apical dendrites that reach and branch within first lamina and several basal dendrites distributed through the SP zone (**a–e**). These early SP neurons together with Cajal-Retzius (C-R) cells of first lamina (**c**) constitute the essential components of the mammalian neocortex early primordial cortical organization. In addition, SP zone Martinotti neurons (**f**), pyramidal neurons (**e**) of the newly establish pyramidal cell plate (PCP), and some undetermined neurons (?) with descending axon (A and G) are also recognized at this age. A rich fiber plexus (**a, b, g, e**) also characterizes the SP zone composed of its axonic collaterals and axon terminals. Some of these neurons axon terminals become the first corticofugal leaving the developing cerebral cortex through a newly established white matter (**a**) zone

with prominent pyramidal-like neurons (Fig. 11.1a–e), a thin band of white matter (Fig. 11.1a, b, f), and a large cellular matrix zone (Fig. 11.1a). Its most distinguishing feature is the presence of numerous large SP zone pyramidal-like neurons (Fig. 11.1b–e). These neurons are characterized by an ascending apical dendrite that reaches and branches within first lamina, by several long basal dendrites and by axonic collateral distributed through the zone and a terminal axon that reaches the white matter and becomes a corticofugal fiber. Their long basal dendrites and axonic collaterals constituted a distinct fibrillary mesh that extends within the SP zone (Fig. 11.1a, b, d, e, g). The SP zone has scattered Martinotti cells with an ascending axon that reaches and branches within first lamina (Fig. 11.1f).

The first lamina is composed of, at least, two distinct types of neurons. Some neurons have short dendrites distributed within first lamina and a descending axon that reaches and branches within the SP zone (Fig. 11.1a, g). The nature of these neurons (also recognized in 25-day-old embryos) remains questionable. These neurons descending axons reach and branch within the SP zone. These axon terminals could reach the white matter and, together with the axons terminals of SP pyramidal-like neurons, become the early corticofugal fibers from the neocortex primordial functional organization. Other larger first lamina neurons, with horizontal dendrites and axonic processes distributed within the lamina, are recognized as C-R cells (Fig. 11.1c). The horizontal axon terminals of some of C-R cells seem to have descending collaterals that reach the SP zone, denoting additional functional interrelationship between both strata.

At this age, the PCP is composed of immature pyramidal neurons with smooth (spineless) apical dendrite attached to first lamina, smooth body, and a short descending axon (Fig. 11.1e). These neurons' apical dendrites are often bifurcate (Fig. 11.1e). The subsequent developmental elongation of these apical dendrites explains the occurrence of dual and parallel apical dendrites among many maturing pyramidal neurons (Figs. 11.4a, 11.6a, 11.7a, 11.8).

At this age, the C–R and the SP zone pyramidal-like and Martinotti cells have morphological interrelationships denoting functional ones as well. Together, they constitute the elements of the mammalian neocortex early primordial functional system. The neocortex few pyramidal neurons are still immature and functionless. Eventually, this primordial functional system, based on the SP pyramidal-like as projecting neurons, will be replaced by a definitive one based on the pyramidal neurons.

11.3 Cat 35-Day-Old Stage

At this age, the SP zone pyramidal-like cells are still the larger and more prominent neurons of the neocortex (Fig. 11.2a). They have long apical dendrites that reach and branch within first lamina and axons that reach the white matter and become corticofugal fibers. Their axons give off several collaterals distributed throughout the zone and an ascending one that reaches and branches within the first lamina (Fig. 11.2a, *arrows*). These large SP zone neurons are not as abundant as they were earlier. On the other hand, the number of pyramidal neurons has significantly increased throughout the developing neocortex PCP (Fig. 11.2a). The morphological differences, including size and location, between SP pyramidal-like and PCP pyramidal neurons are readily appreciated in some Golgi preparations (Fig. 11.2a). The first lamina C-R cells have already developed their typical long horizontal dendrites and axonic processes (Fig. 11.2b, c). These neurons tend to occupy the upper region of the lamina and their processes are intermingled with the horizontal axon terminals of primordial corticipetal fibers (Fig. 11.2c). See also Chapters 3 and 5.

The developing neocortex PCP is already eight to ten cells thick. All pyramidal neurons are still immature with apical dendrites, of different length and often bifurcated, anchored to first lamina by terminal dendritic bouquets (Fig. 11.1a). They all have smooth and spineless apical dendrites, smooth somas, and short descending axons with terminal growth cones that are approaching the white matter (Fig. 11.2a). Ascending primordial corticipetal fibers cross unbranched the PCP, reach the first lamina, and become long horizontal terminals (Fig. 11.2c). They represent the only afferent fiber system reaching the cat neocortex at this age.

Because of the developmental significance of this developmental stage, a montage of camera lucida drawings from Golgi preparations of several 35-d-o fetuses has been prepared (Fig. 11.3). This developmental stage represents the mammalian neocortex primordial morphological and functional organization. The drawing summarizes, at scale, the basic cytoarchitectural organization of the cat motor cortex, at this age (Fig. 11.3). The drawing illustrates comparatively (at similar scale) the size of the essential neuronal types of the cat motor cortex as well as their location, distribution, morphological, and possibly functional interrelationships. Primordial corticipetal fibers from the white matter ascend unbranched through the PCP, reach

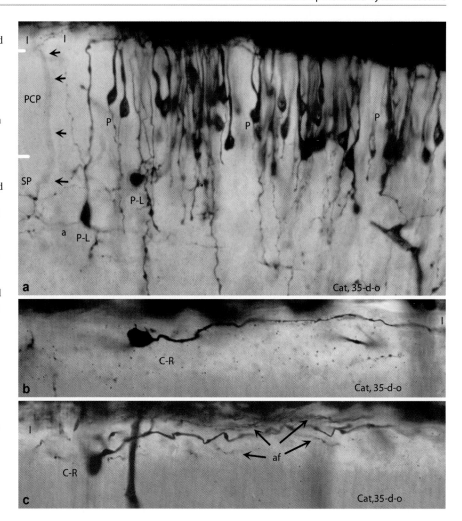

Fig. 11.2 Montage of photomicrographs, from rapid Golgi preparations, of the motor cortex of 35-day-old cat fetuses, illustrating a general view of (**a**) of the cortex cytoarchitectural organization and composition and two (**b, c**) detailed views of the morphologic features of Cajal-Retzius (C-R) cells. A: Illustrates comparatively the numerous undifferentiated newly arrived pyramidal (P) neurons of the pyramidal cell plate (PCP) with ascending smooth apical dendrites branching within first lamina with the large SP pyramidal-like (P-L) neurons with ascending apical dendrite and axonic (a) collateral (*arrows*) that reach and branch within the first lamina. (**b** and **c**) Illustrate the long horizontal dendrites of first (I) lamina Cajal-Retzius (C-R) cells intermingled with the axonic terminals (at) of primordial corticipetal fibers representing the first arriving extrinsic (extracortical) fibers at the developing cerebral cortex

the first lamina, and become long horizontal fibers, which intermingle with the C-R cell processes (Fig. 11.3). Many fibers of the white matter have terminal growth cones advancing on opposite directions representing either arriving corticipetal or departing corticofugal fibers (Fig. 11.3). At this age, most of the corticofugal fibers leaving the neocortex are the axon terminals of the SP pyramidal-like neurons (Fig. 11.3).

11.4 Cat 40-Day-Old Stage

At this age, the cat motor cortex is at a crucial stage characterized by the starting ascending functional maturation of the PCP lower and older pyramidal neurons with the establishment of the first P1 pyramidal cell stratum (Fig. 11.4a). This stage may be comparable to the human motor cortex at 15 weeks of gestation (see Chapter 3). The PCP deepest and older pyramidal neurons have started to develop basal, lateral dendrites and a few proximal apical dendritic spines (Fig. 11.4a). Some P1 stratum maturing neurons have bifurcated apical dendrites attached to first lamina (Fig. 11.4a). The remaining PCP pyramidal neurons remain immature with smooth and spineless apical dendrites anchored to first lamina, smooth somas without basal dendrites, and short descending growing axons (Fig. 11.4a). It is important to point out that an intrinsic capillary plexus has already developed throughout the P1 pyramidal cell stratum supporting their early functional maturation (Fig. 11.4a, *arrows*).

The functional maturation of C-R cells has progressed and their long horizontal dendrites and axonic collaterals have extended throughout the first lamina upper region intermingled with afferent fibers terminals (Fig. 11.4b). The number of horizontal fiber terminals has also increased through the first lamina.

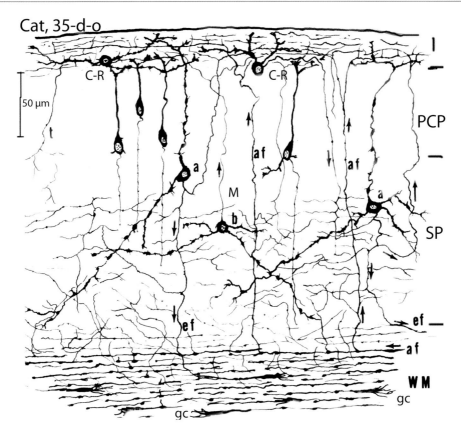

Fig. 11.3 A camera lucida drawing, from rapid Golgi preparations, summarizes comparatively, at similar scale, the overall cytoarchitectural organization and neuronal composition of the cerebral cortex of 35-day-old cat fetuses. Primordial corticipetal fibers (af), from the white matter (WM), ascend unbranched up to the first lamina and terminate into long horizontal processes, the SP plate pyramidal neurons have ascending dendrites and azonic collaterals that reach and branch into first lamina and axon terminal (ef) that enter into the underlying white matter and become the first corticofugal fibers leaving the developing cortex. The pyramidal cell plate (PCP) pyramidal neurons are still immature with smooth apical dendrites and somas and descending axon that start to reach the underlying white matter. The underlying white matter fibers have fibers with terminal growth cones advancing in different directions representing corticipetal and/or corticofugal fibers. Other features of the drawing are self-explanatory

From this gestational age onward, the SP zone, due to the PCP expansion, has descended progressively within the neocortex. Additional deep Golgi preparations will be necessary to explore its further development.

11.5 Cat 45-Day-Old Stage

At this age, some SP pyramidal-like and Martinotti neurons have started to lose their functional attachment with the first lamina (Fig. 11.5a, b). Some of the disconnected apical dendrites of SP pyramidal-like neurons show terminal beadings, indicating early signs of dendritic degeneration (Fig. 11.5a, b). Also, some deep SP Martinotti cells are starting to regress their axonic terminals (Fig. 11.5a, b). This cat developmental stage is comparable to the human 20-w-o stage (see Chapter 3).

The functional maturation of pyramidal neurons of stratum P1 has advanced. Their basal and lateral dendrites are longer and have more dendritic spines (Fig. 11.5b). In addition, some PCP pyramidal neurons, above stratum P1, have started to develop short basal dendrites and dendritic spines denoting their starting ascending functional maturation and the establishment of the second P2 pyramidal cell stratum (Fig. 11.5a, b). All PCP pyramidal neurons above the newly established P2 stratum remain immature and spineless.

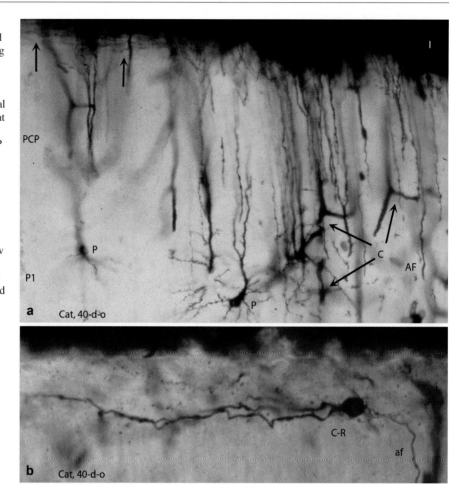

Fig. 11.4 Montage with two photomicrographs, from rapid Golgi preparations, illustrating (**a**) a general view of the cerebral cortex and a closer view (**b**) of a Cajal-Retzius (C-R) cell with long horizontal processes, from 40-day-old cat fetuses. The deep and older pyramidal neurons of the PCP have started to develop short basal and lateral dendrites denoting their starting ascending functional maturation and have established the cortex first pyramidal cell P1 functional stratum. (**a**) Noticed that a few capillaries (*arrows*), from the cortex intrinsic microvascular system, are already established throughout this functional stratum, paralleling its pyramidal neurons ascending maturation. Some axonic terminals (*arrows*) from C-R cells are also recognized in first lamina between silver deposits. (**b**) Detail of a C-R neuron with long horizontal dendrites intermingled with afferent fiber terminals from ascending (AF) primordial corticipetal fibers

A camera lucida drawing, from rapid Golgi preparations of 45-d-o cat fetuses, recapitulates the cat motor cortex cytoarchitectural development at this age, the establishments of P1 and P2 pyramidal cells strata, and the starting regression of the SP neurons' functional contacts with first lamina (Fig. 11.5b). The cat motor cortex thickness has increased due to the PCP expansion, which is already 300 μm thick. The PCP expansion has displaced downward the SP zone, which is now located 600 μm below pial surface (Fig. 11.5b). The SP zone neurons, still quite prominent in the motor cortex, have started to lose their first lamina functional contacts. The drawing demonstrates, comparatively, the size, morphology, location, distribution, and interrelationships among the main neurons of the cat motor cortex, at this age (Fig. 11.5). Other camera lucida drawing features are self-explanatory (Fig. 11.5).

11.6 Cat 50-Day-Old Stage

At this age, the cat motor cortex is characterized by the extraordinary functional maturation already achieved by the pyramidal neurons of P1 stratum and the starting functional maturation of pyramidal neurons of stratum P2 (Fig. 11.6a). At this level, some pyramidal neurons have started to develop

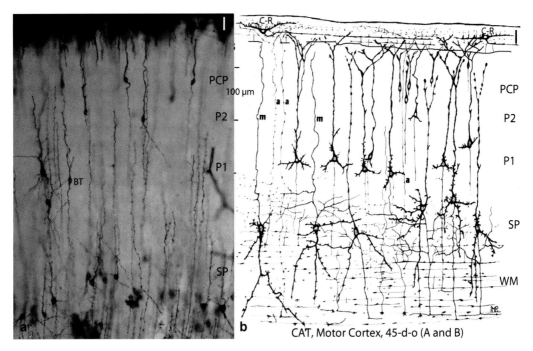

Fig. 11.5 Composite figure with a photomicrographs (**a**) and a camera lucida (**b**), from a rapid Golgi preparations, illustrating its basic cytoarchitectural organization and composition of the motor cortex of 45-day-old cat fetuses. At this age, the cat motor cortex is characterized by the pyramidal cell plate (PCP) expansion, the downward displacement of the subplate (SP) zone, the SP zone pyramidal-like and Martinotti neurons starting functional detachment from first lamina, and the starting retraction of its ascending processes. The pyramidal neurons of P1 stratum have longer and more numerous basal and lateral dendrites and the pyramidal neurons of P2 stratum are starting to develop basal dendrites denoting that they are starting their ascending functional maturation. The cytoarchitecture of the cat motor cortex at this age is comparable with the motor cortex of 20-week-old human fetuses (See Chapter 2). Other features of both (**a**) and (**b**) are self-explanatory

basal dendrites and a few dendritic spines (Fig. 11.6a). The formation of basal dendrites and the appearance of dendritic spine on these neurons imply the level reached by ascending corticipetal fibers and the establishment of functional contacts with them. The PCP upper pyramidal neurons remain immature and undifferentiated with smooth apical dendrites and somas. This new ascending corticipetal fiber system is probably of thalamic origin. Its ascending fibers have reached the pyramidal cell P2 stratum level at this age. The axons of pyramidal neurons from P1 and P2 strata have reached the white matter and become the early corticofugal fibers of the mammalian neocortex definitive functional system.

At this age, most of the SP zone pyramidal-like and Martinotti neurons are already disconnected from the first lamina and have assumed the morphologic features of deep polymorphous neurons (Fig. 11.6a, b). They are scattered, among the increasing number of fibers throughout the SP zone. The SP zone can no longer be distinguished from the increasing white matter zone. The depth of these deep polymorphous neurons ranges from 600 to 700 μm from pial surface (Fig. 11.6b).

11.7 Cat 55-Day-Old Stage

A montage of camera lucida drawings, from rapid Golgi preparations, illustrates comparatively and to scale, the cat motor cortex cytoarchitectural organization and

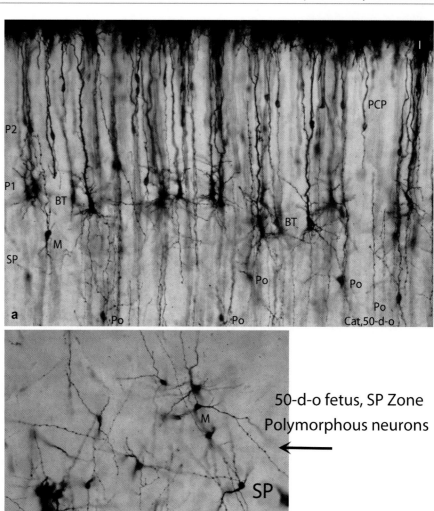

Fig. 11.6 Composite figure of photomicrographs illustrating the cat developing motor cortex basic cytoarchitectural organization and composition, from rapid Golgi preparations of 50-day-old fetuses. (**a**) Illustrates the degree of the ascending functional maturation and distinct stratification of pyramidal neurons of P1 stratum, the starting maturation of P2 stratum pyramidal neurons, and the still undifferentiated pyramidal neurons of the remaining PCP zone. Some of the deep SP zone neurons are starting to assume the morphologic feature of deep polymorphous (Po) neurons. (**b**) Detailed view of cortex SP zone showing the morphologic features of deep polymorphous and Martinotti (M) neurons already functionally disconnected from the first lamina

composition, at this age (Fig. 11.7a). The developmental maturation achieved by the pyramidal neurons of stratum P1, located at 500 μm from pial surface, is remarkable. The pyramidal neurons of P2 stratum have already developed several basal dendrites and those of P3 stratum have started their ascending functional maturation by developing short basal dendrites (Fig. 11.7a). All the pyramidal neurons above these two strata are still functionally undifferentiated. Most SP zone neurons are functionally disconnected from the first lamina and transformed into deep polymorphous neurons scattered, at depth ranging from 600 to 800 μm from pial surface, among the increasing number of white matter fibers (Fig. 11.7a).

11.8 Newborn Cat Motor Cortex

The newborn cat motor cortex is characterized by the functional maturation of an additional P4 pyramidal cell stratum to those P1, P2, P3, and P4, already

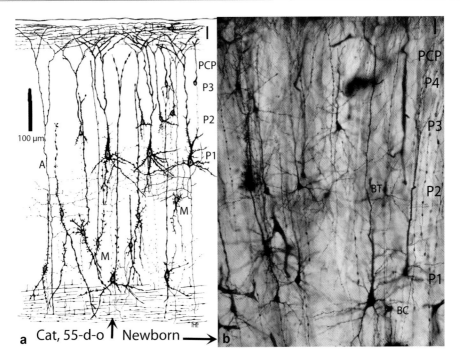

Fig. 11.7 Composite figure with (**a**) a camera lucida drawing from the developing motor cortex from 55-day-old cat fetuses and (**b**) a photomicrograph from rapid Golgi preparation of a newborn cat motor cortex. (**a**) The camera lucida drawing illustrates the basic cytoarchitectural organization and composition of developing cat motor cortex at this age. It includes the advanced ascending maturation of the pyramidal neurons of P1 and P2 strata, the starting maturation of pyramidal neurons of P3 stratum, and the undifferentiated pyramidal neurons of the remaining pyramidal cell plate (PCO). Al, PCP pyramidal neurons are functionally anchored to the first lamina (I). All SP zone neurons are functionally disconnected from first lamina and have assumed the morphological features of deep polymorphous (interstitial) neurons. (**b**) A rapid Golgi photomicrograph illustrating the basic cytoarchitectural organization and composition of a newborn cat motor cortex, including the ongoing ascending functional maturation of P1, P2, P3, and P4 strata pyramidal neurons, the presence of some inhibitory basket (BC) cells of P1 stratum, the residual PCP still undifferentiated, and several ascending fibers with rosaries representing degenerating radial glial fibers

established (Fig. 11.7b). The functional maturation of the new P4 pyramidal cell stratum is paralleled by the ascending maturation of its corresponding intrinsic microvascular, protoplasmic astrocytes, and excitatory–inhibitory neuronal systems. The newborn cat motor cortex is further characterized by the presence, beneath the first lamina, of a band of still undifferentiated PCP (Fig. 11.7b).

Two additional camera lucida drawings, from rapid Golgi preparations, are submitted for general information and to recapitulate the basic developmental observations presented. The first one illustrates, comparatively, the developmental histories of the motor cortex essential neuronal systems, namely, the SP zone pyramidal-like and Martinotti neurons and the stratum P1 pyramidal neurons (Fig. 11.8). The evolving morphology, size, location, and distribution of these three neuronal types of the cat motor cortex are reproduced through the 27th-, 30th-, 35th-, 40th-, 45th-, and 50-d-o fetuses and the newborn (Marín-Padilla 1972). The SP zone pyramidal-like and Martinotti neurons undergo a progressive developmental regression and eventual transformation into deep polymorphous (interstitial) neurons. The regressive developmental histories of the SP zone neurons contrast sharply with that of pyramidal neurons characterized by increasing morphological and functional complexities (Fig. 11.8). The pyramidal neuron represents a mammalian innovation that distinguishes their cerebral cortex from that of any other vertebrate. The number of pyramidal cell strata established in the

Fig. 11.8 Montage of camera lucida drawings, from rapid Golgi preparations, illustrating the prenatal developmental histories and morphological transformations of the cat Martinotti cells, gray matter pyramidal neurons, and SP zone pyramidal-like neurons through the 27th-, 30th-, 40th-, 45th-, and 50th-day-old fetuses and from a newborn. The newborn cat neurons depth (from the cortex pial surface) is also recorded. Other features of the drawings are self-explanatory. (Modified from Marín-Padilla 1972)

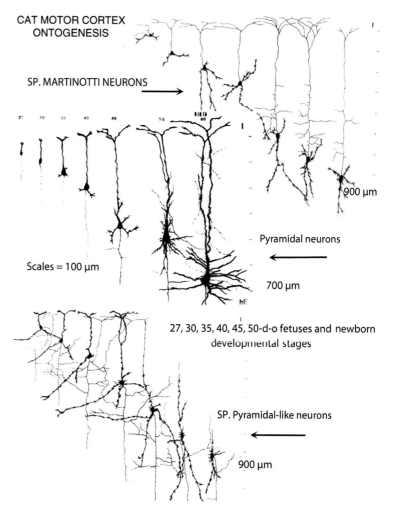

developing mammalian neocortex varies and, in essence, reflects the animal motor capabilities (see Chapters 3 and 9). The essence of the developing mammalian neocortex could be thought as the transformation of a primordial functional organization, coordinated by the SP zone neurons in conjunction with C-R cells, into a mammalian functional system coordinated by various pyramidal cell strata, also in conjunction with C-R cells. As pyramidal neurons assume the mammalian neocortex final and definitive functional control, the SP zone neurons lose their first lamina functional contacts and regress progressively. Their persistence in the cerebral cortex as deep interstitial neurons suggests that they have retained some – so far undetermined – functional activity. Perhaps, they represent relay interneurons between thalamic nuclei and the neocortex gray matter. Their possible functional role remains undetermined and further investigation is needed.

The second drawing illustrates, comparatively and to scale, general views of the cat developing cerebral cortex through the 25th-, 30th-, 35th-, and 40th-day of gestation, respectively (Fig. 11.9). They reproduce the cat developing neocortex extraordinary neuronal and fibrillary expansion and increasing complexity during its first

Fig. 11.9 Montage of camera lucida drawings, from rapid Golgi preparations, showing comparatively, and at scale, general views of the developing cerebral cortex of cat through 25-, 30-, 35-, and 40-day old fetuses. The proximal (ventral) to distal (dorsal) developmental maturation gradients of the different cortical strata and of their composition are also illustrated. The early ongoing prenatal development and maturation of the cat cerebral cortex can be easily appreciated through these drawings

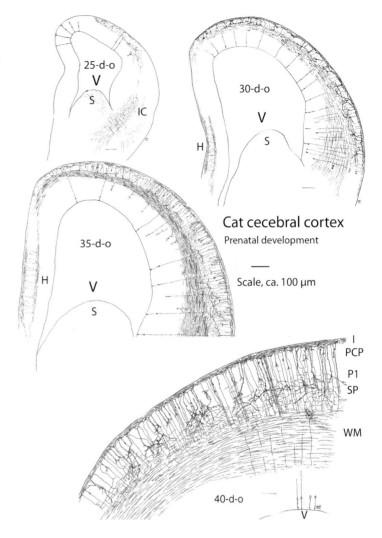

40 days of life. Also, they illustrate the development of the neocortex principal functional strata, the first lamina persistence, the increasing thickness of the PCP, and the early maturation of the first P1 pyramidal cell stratum (Fig. 11.9). The drawings demonstrate the neocortex ventral-to-dorsal (proximal-to-distal) developmental maturation gradient. Such that the neocortex proximal regions are developmentally more advance than the distal ones (Fig. 11.9). While the developing PCP is already established through proximal cortical regions, the distal regions are still at the primordial plexiform developmental stage, in both 30- and 35-d-o embryos (Fig. 11.9; see also Fig. 2.3a of Chapter 2). The 100 μm scale is roughly applicable for all drawings.

It is important to emphasize that the prenatal neuronal development of the cat motor cortex corroborates that observed in humans (see Chapter 3). The only significant as well as essential difference between them is in the number of pyramidal cell strata established on their respective cerebral cortex. Only four pyramidal cell strata are established in the cat motor cortex, while six are established in the human motor cortex. These differences, essentially, reflect the motor capabilities of each species.

References

Colby DE (1970) Induced estrus and time pregnancy in cats. Lab Animal Care 20:1075–1080

Luskin MI, Shatz CJ (1985) Studies of the earliest generated cells of the cat's visual cortex: cogeneration of subplate and marginal zones. J Neurosci 5:1062–1075

Marín-Padilla M (1971) Early prenatal ontogenesis of the cerebral cortex (neocortex) of the cat (Felix domestica). A Golgi study. I. The primordial neocortical organization. Zeitschrift für Anatomie Entwicklungeschichte 134:117–145

Marín-Padilla M (1972) Prenatal ontogenetic history of the principal neurons of the neocortex of the cat (Felix domestica). A Golgi study. II. Developmental differences and their significance. Zeitschrift für Anatomie Entwicklungeschichte 136:125–142

Marín-Padilla M (1978) Dual origin of the mammalian neocortex and evolution of the cortical plate. Anat Embryol 152: 109–126

Marín-Padilla M (1998) Cajal-Retzius cells and the development of the neocortex. Trends Neurosci 21:64–71

The Rapid Golgi Reaction: A Personal Quest

In 1873, Camilo Golgi, Professor of Pavia, introduced his "reazione nera" or "black reaction" (Golgi 1873). It is a simple neurohistological procedure that consists on the fixation and staining of tissue blocks (about 9×7×3 mm). The blocks are fixed in a potassium dichromate and osmic acid solution and staining in a silver nitrate solution. The already fixed and stained blocks are cut (free hand with a razor blade) perpendicular to the gyrus long axis, into 150–250 μm thick slices, which are subsequently dehydrated in a 95% alcohol solution, cleared in oil of cloves, mounted sequentially in a large glass slide, and covered, individually, with a drop of Damar-xylene solution. After drying, the slices are ready for microscopic study. These thick preparations allow the visualization of neurons (often in their entirety), axonic terminals, synaptic profiles, glial elements, and the sprouts of growing blood capillaries of the developing brain as well as their overall interrelationships. Additional and more detailed analyses and descriptions of the procedure and some historical background can be found in Cajal (1911); Valverde (1970); and Marín-Padilla (1971, 1978a, 1990, 1992) works.

It is possible to state that our basic understanding of the structural organization of most regions of the nervous system, of invertebrates, as well as vertebrates stands on Golgi observations. It is therefore perplexing that this classic procedure, unfortunately, seldom used nowadays, has remained so poorly understood and surrounded by misconceptions and criticisms. Despite controversies, the technique continues to be unmatched for exploring the cytoarchitecture of the developing mammalian neocortex, including humans, as well as of many other regions of the nervous system of vertebrates and invertebrates. The procedure is capable of staining the fundamental neuronal, fibrillar, synaptic profiles, microvascular, and glial components of the central nervous system. In my opinion, its clinical use should be encouraged because the structural organization of many regions of the human brain remains unknown and, also because, the neuronal organization of most major neurological and psychological disorders, such as epilepsy, cerebral palsy, dyslexia, autism, mental retardation, early Alzheimer, continued to be unknown and essentially unexplored.

My first contact with the procedure occurred, in 1967, at the Cajal Institute of Madrid, Spain. I was the recipient of an NIH (National Institute of Health, USA) Neurohistology Fellowship, to study Cajal's old Golgi preparations, kept at the institute museum. The idea was to place my eyes on Cajal's and learn firsthand, from his old Golgi preparations, the procedure's features, applicability, capabilities, and limitations. Also, trying to discern why was Cajal able to see so clearly and so much more than anyone else since. Cajal described Golgi observations in these words: "An unexpected spectacle! On a perfectly translucid yellow background there appeared scattered black filaments, smooth and narrow or spiny and thick; black bodies, triangular, stellate, fusiform! One would have thought they were ink drawings on transparent Japanese paper. They trouble the eye. Everything looked simple, clear, unconfused. Nothing remained but to interpret. One merely had to look and note" (Cajal 1911). And, indeed, he looked, interpreted, and took innumerable notes. I concur with his thoughts and words, and share his enthusiasm with the procedure. My first human cerebral cortex publication was based on Cajal's old Golgi preparations (Marín-Padilla 1967).

After the incomparable experience of studying Cajal's Golgi preparations and convinced of the procedure's possibilities, I started, back in Dartmouth, to manipulate its rapid variant for the study of the developing cerebral cortex of humans, cats, hamsters, mice, and rats embryos and continued, using it, throughout my academic years (1967–2003). My Golgi studies, based on more than 4,000 preparations, are scattered through 28 abstracts, 42 peer review papers, and 12 book chapters.

An examination of the procedure's essential features, qualifications, revealing potentials, peculiarities, and technical quandaries, from a personal perspective, might provide a better understanding of it, elucidate some of the misconceptions and, perhaps, encourage its usage. To start is necessary to accept that the Golgi reaction is, indeed, a capricious procedure and learning to manipulate it successfully represents a long and personal quest. The Golgi procedure's basic features include technical simplicity, flexibility, adaptability, universal applicability, staining selectivity, and the use of thick preparations, permitting three-dimensional views of the nervous organization.

First, it must be recognized that the procedure does not represent a method, in the strict sense, but a reaction characterized by extraordinary flexibility and adaptability. A fact already recognized by Golgi, who remarked that, the secret of his "black reaction" rested in "provando a reprovando," trying and trying again. Consequently, the term "Golgi reaction" best reflects the procedure's nature and Golgi's original description of it. Therefore, it should replace the commonly used term of "Golgi method." Every investigator, who has used it, has introduced some modifications, some of which have achieved notoriety, such as the rapid Golgi for unfixed tissue and the Golgi-Cox and Golgi-Kopsch, for previously fixed tissue. Its rapid variant, favored by Cajal and myself, is simple and ideal for studying the developing nervous tissue.

Second, both the Golgi reaction fixation and impregnation times are critical and, possibly, specific for each region of the nervous tissue, each animal species, and each developmental age. Only by – trying and trying again – can these critical times be preestablished for any study using the procedure. For example, to study the human cerebral cortex first lamina and avoid superficial silver deposits, both fixation and impregnation times were reduced to only a few hours each (Marín-Padilla 1990). While layer I may be ideally fixed and stained, deeper cortical regions were not. In contrast, to study deeper cortical regions, fixation and impregnation times should be increased, however, layer I cytoarchitecture will be buried under silver deposits. Some of the misconceptions regarding the mammalian neocortex first lamina, which due to this predicament are invariably obscured by silver precipitates, a fact already recognized by Cajal. He recurred to encase the brain section I coagulated blood to minimize surface silver precipitates. Perhaps, this simple fact explains why this important cortical lamina has not been adequately visualized and studied.

Third, the thick (150–250 μm) and nearly transparent Golgi preparations (rapid variant) permit unparallel 3-D views of the nervous tissue cytoarchitectural organization. They also provide unique 3-D vistas of the interrelationships among neurons, axonic terminals, dendritic branches, synaptic profiles, glial cells, and blood capillaries. The microscopic study of such preparations permits the exploration of at least three successive levels (upper, middle, and lower) on each section. Moreover, already fixed and stained tissue blocks can be cut perpendicular, parallel, and/or tangential to the gyrus axis, depending on the objective, supplying additional 3-D views, of the same region, from different perspectives. No other available neurohistological procedure offers such multiple 3-D vistas of the nervous tissue structural organization. Due to its remarkable flexibility, the Golgi reaction applicability is virtually universal in the study of the nervous system of any species, any region, and/or any developmental age. From the insect to the human eye, from the fish to the human cerebellum, and from the amphibian to the human cerebral cortex, all have been successfully explored and better understood, with the usage of the procedure.

Fourth, for best results (rapid Golgi variant), is indispensable and strongly recommended to utilize fresh, unfixed, and unspoiled brain tissue, which is not always possible. The human brain neuronal cytoarchitecture deteriorated rapidly, after 2–4 h postmortem, fragmentation of cortical elements becomes quite obvious and their correct interpretation increasingly difficult to make. For clinical studies, the gathering of fresh human brain tissue requires the collaboration and understanding of clinicians, nurses, families, and hospital personal, which are not always easy to achieve.

Fifth, for a single investigator, the usage of the procedure, for clinical studies, has its limitations. The number of tissue blocks that can be processed, at any time, is limited. The cutting, clearing, and mounting of sections are time consuming, consequently, any uncut tissue block remaining in either fixative or silver solution will be ruined if it is not processed in due time. Ideally, when a fresh postmortem human brain becomes available, there should be several investigators available and ready to process as many blocks as possible and, after completing the studies, compare results. Presently, this ideal situation is unfeasible. Despite these constrains, as Director of the Pediatrics Autopsy Service (Dartmouth Hitchcock Medical Center), I was able, with the participation and approval of every one, to process and collect nearly 5,000 rapid Golgi preparations of the developing cerebral cortex of humans, cats, hamsters, and mice embryos. In this collection, some sections are well stained, other are poorly so, while still others are only partially stained, Any Golgi preparation, good or bad, has always something to offer to the patient's eye.

The shared number of sections made and the number of years dedicated to their study compensate for those shortcomings. Despite these limitations, the amount of information contained on any Golgi preparations is enormous and the needs of returning to them, to reconfirm findings and/or to search for new ones, are mandatory. A single viewing of a Golgi preparation is invariably insufficient. Because Golgi preparations are practically everlasting, one can return to study them as many times as is necessary. Even the dendritic spines delicate structure is still clearly visible in Cajal's old Golgi preparations, already 110 years old. Golgi preparations are like indestructible fragments of amber with blackish incrustations of biological vestiges. The search for biological vestiges, through Golgi preparations, is a challenging scientific proposal. Both, the preparation and the study of Golgi preparations must be a lifelong commitment; shorter amount of time will not do.

Sixth, a significant drawback of the procedure is the common occurrence of silver precipitates, which obstruct the view of stained elements, must be avoided at all cost. In order to increase the number of stained neurons, it has been customary to either increase the silver impregnation time and/or to reimpregnate the preparations. Either practice may render some preparation useless, their use should be discouraged. An important rule to remember is that it is preferable to study understained Golgi preparations than overstained one. The fewer elements stained will be seen clearly, silver deposits will be reduced, and the thickness of the entire preparation will be transparent and without interferences. It is, therefore, advisable to prepare as many understained sections as possible, since their study will surely compensate for the shortcoming.

Seventh, technically the Golgi reaction is simple, easy to learn, and economical. Some manual dexterity is required because best results are obtained by cutting, already fixed and stained blocks, free hand (with a razor blade) minimizing unnecessary tissue damaged. The recommended encasing of each tissue block in a hot paraffin turret requires drying it previously. Both, the hot paraffin and the dryness will damage the nervous tissue'sdelicate structure. Any other mechanism to secure the tissue block, for cutting, is equally damaging. In a good-quality Golgi preparation, stained neurons, dendrites, axonic terminals, synaptic sites, glial cells, and capillaries are viewed intermingle with each other. The best Golgi preparations are those with fewer elements stained. Cajal used these words in describing the reaction and advise the observer: "Thirty five years of using, nearly exclusively, the Golgi method have persuaded me of the magnitude of possible causes for error and have made me suspicious y very prudent of the presence of unusual structural dispositions." Cajal expresses the responsibility of the observer in interpreting correctly the structural organization of the nervous tissue, using Golgi preparations, and the need to reconfirm observations by repeatedly reviewing the same one.

Frequently expressed criticisms of the Golgi reaction are that it only stains a few neurons and that fails to stain all neuronal types. Neither criticism is entirely correct. It is unrealistic to expect that all neurons and/or neuronal types should be stained on each Golgi preparation; in such a case, the preparation will be totally black and useless. For example, the cutting of a single tissue block (ca. 3 mm thick) could yield around 20 (250 μm thick) already stained consecutive sections. The preparation of five tissue blocks per specimen (I have used regularly) will result in 100 such sections and each one will offer, at least, three levels (upper, middle, and lower) for inspection. Therefore, a single brain specimen could yield 300 different levels for inspection, which surely will have several stained neurons, different neuronal types, as well as many other stained elements. The problem is that the study of the 300 different levels, for each single specimen, will require not hours or days but weeks and months and, in some cases, even years. Discouragement and lack of sufficient time have kept investigators away from the procedure. Many of my students fascinated, in my classes, after seeing the beauty of some Golgi preparations, wanted to work in my laboratory, once they realize the degree of commitment involved decide to move away. Mastering the Golgi reaction requires personal commitment, perseverance, veracity, and must be equated to a personal and enduring quest.

Two other negative claims, expressed about the Golgi reaction, are that it is unreliable and that the observations are irreproducible. Both claims are mistaken and ignore the nature of the nervous system itself. The structural organization of any particular area of the nervous system is, in itself, unique and irreproducible. Despite obvious similarities, the cytoarchitectural organization of neurons, fibers, synaptic profile, glial, and blood capillaries, on any Golgi section, are unique for that location and age and different from those in preceding and/or following cuts. Hence, any Golgi section is, by nature, irreproducible. To explore, document, and record well-stained, although fragmentary, elements and ignore those that are not well stained or unstained, have been my motto through the years. In a good-quality rapid Golgi preparation, any stained element, regardless of quantities, will be clear, reliable, and irrefutable. The strategy rested in making as many Golgi preparations as possible, foreseeing that eventually one of them will have what one is searching for. The study of a Golgi section may be comparable to exploring a forest (a simile often used by Cajal). Exploring a forest will yield, at every step, a new vista, the searching for new vistas will always be exhilarating and the return to the original one for additional detailed observations will certainly be compulsory.

It is important to recognize that since the amount of information in any Golgi preparation may be variable and overwhelming, the use of camera lucida drawings is obligatory and required. The location, size, morphology, and interrelations of stained elements can only be reproduced and illustrated using camera lucida drawings. Photomicrography alone is inadequate for a variety of reasons. A photomicrograph will depict all stained elements, desirable and undesirable, within a single plane, which often is rather confusing. The superposition of all stained elements, within it, makes the identification of any particular one difficult. A photograph also fails to convey the depth of stained elements and their 3-D interrelationships. Similarly, computer reconstructions of Golgi preparations are inadequate and rather confusing. They reproduced everything stained, including undesirable silver deposits, blood capillaries, and glial elements. Elements stained through the upper level overlap, intermingle, and obscure those in the middle and lower ones. The result is a confusing arrangement of stained elements impossible to unravel. The few 3-D computer reconstructions of a Golgi preparation, that I have helped in making and have the opportunity of reviewing, were extremely complex and difficult to understand. Only the direct microscopic observation of a Golgi preparation, by moving the objective up or down, permits a general view of all stained elements through the various cortical levels, determine their morphology, location, depth, size, and interrelationships as well as the possibility of distinguishing each one as a separate entity. Reproducing this 3-D cytoarchitectural organization using photomicrography will be utterly confusing as all stained elements overlap with each other. Only the use of a camera lucida attached to the microscope permits the reproduction of this complex 3-D organization.

Studying Golgi preparations, rough camera lucida drawings are regularly made, depicting, at similar magnification, the morphology, size, and location of any selected structure. When many such drawings become available, from each case, they are restudied searching for the repetition of particular structures. Once a specific element is selected from these drawings, it will be reviewed microscopically for additional morphological details and a final camera lucida drawing, at higher magnification, will be made of it and its neighboring elements. Montages of selected camera lucida drawings are made to illustrate, at same magnification, all the element visualized. Both montages of camera lucida drawings and corresponding photomicrographs (when possible) are prepared to accompany the Golgi description of each case. These serial processes are repeated until the structural organization of the cortical region under study starts to be better understood. The montages of camera lucida drawings, from Golgi preparations, available throughout the scientific literature, reproduce what cannot be illustrated by using photography alone. In good-quality Golgi preparation, selected microphotography is not only possible but also outstanding, as those accompanying the present monograph will attest.

A significant drawback in the use of the Golgi reaction is its inadequacy to study the adult central nervous tissue. The presence of myelin seems to interfere with the reaction, making the study of an adult brain very difficult, fragmented, and unreliable.

In addition to the numerous examples found through the literature, a brief account of a few personal observations obtained with the Golgi reaction, rapid variant, should substantiate and corroborate the above-expressed assertions and comments, and hopefully encourage the continued usage of this classic procedure.

Some fundamental observations concerning the developing mammalian neocortex are based on Golgi studies. They include: (a) the establishment of its dual origin, later corroborated and unanimously accepted today; (b) that its primordial (embryonic) cytoarchitectural organization shares features with the amphibian and reptilian cerebral cortices; (c) that its pyramidal cell plate is a mammalian innovation, which is formed and expanded within the primordial neocortical organization, establishing, simultaneously, its first lamina and the subplate zone; and, (d) that its pyramidal neuron ascending morphological and functional maturations represent unique mammalian innovations (Marín-Padilla 1970, 1971, 1972a, b, 1978a, b, 1985a, 1990, 1992; Marín-Padilla and Marín-Padilla 1982).

The Cajal-Retzius multipolar dendritic morphology, tangential distribution of its horizontal axonic collaterals, and multidirectional distribution of its terminal axonic processes have all been demonstrated with Golgi preparations, of the human motor cortex, cut perpendicular, parallel, and tangential to the precentral gyrus long axis (Marín-Padilla 1990).

The fact that the neocortex pyramidal neurons retain their early functional attachment to first lamina and the location (depth) of their bodies throughout their anatomical and functional maturations are also based on Golgi studies (Marín-Padilla 1992). Early, the pyramidal neuron apical dendrite elongates, anatomically, without losing either anchorage, to accommodate the subsequent arrival of new pyramidal neurons. Later, the pyramidal neuron apical dendrite elongates, physiologically, by the ascending addition of functionally active segments for the formation of synaptic contacts (Marín-Padilla 1992).

The original description of the basket cell of the human motor cortex and their 3-D structural organization of its terminal pericellular nests are also based on Golgi studies (Marín-Padilla 1969). Moreover, the basket cell unique 3-D spatial distribution within thin (45–55 µm thick) and rectangular (1,000 µm^2) tissue slabs within the motor cortex, based on Golgi studies, was later confirmed by computer reconstructions (Marín-Padilla 1974; Marín-Padilla and Stibitz 1974).

Dendritic spines morphological abnormalities were first described in Golgi studies of the motor cortex of children with chromosomal trisomies (Marín-Padilla, 1972a, b, 1974, 1968, 1978a, b). Subsequently, dendritic spines anomalies have been described in other neurological and psychological disorders.

Original Golgi studies of the outcome of perinatal brain damage have demonstrated a variety of post-injury cytoarchitectural alterations, involving neurons from the damaged site and from adjacent regions functionally interconnected with them (Marín-Padilla 1996, 1997, 1999). Post-injury cytoarchitectural alterations are believed to play a crucial role in the pathogenesis of ensuing neurological and psychological disabilities in children that survive perinatal brain damage (Marín-Padilla 2010)

The early perforation of the pial surface by capillaries from the pial plexus and their subsequent participation in the establishment of the neocortex intrinsic microvascular system are also based on Golgi and electron-microscopic studies (Marín-Padilla 1985b). Golgi studies have also contributed to a better understanding of origin and prenatal development of both first lamina special glia cells and gray matter protoplasmic astrocytes (Marín-Padilla 1995).

A new cytoarchitectonic conception concerning the mammalian neocortex ascending stratification during their phylogenetic and ontogenetic evolutions is also based on Golgi studies (Marín-Padilla 1998). The new theory proposes that the number of ascending pyramidal cell strata established in the neocortex (motor region) reflects the animal motor needs that it increases paralleling the motor dexterity of each mammalian species. These Golgi studies have challenged the universally held view that the mammalian neocortex has six (or seven) descending basic laminations, namely, layers I, II, III, IV, V, and VI. In order to accommodate the six-lamina criteria, some cortical layers have been arbitrarily condensed, in lower mammalian species, into a single one: layer II-III-IV in the European hedgehog and layer II-III in the mouse neocortex. Conversely, in the neocortex of primates, including man, layer III has been arbitrarily duplicate and even triplicate (IIIa, IIIb, IIIc) and layer II duplicated (IIa, IIb). In general terms, the new cytoarchitectonic theory proposes that insectivores need only two basic pyramidal cell strata to fulfill all their motor needs, rodents three, carnivores four, primates five, and human six (Marín-Padilla 1998).

It is hoped that this brief account of the Golgi reaction accomplishments, universal applicability, and capabilities may encourage young neuroscientists to use it.

References

Ramón y Cajal S (1911) Histologie du Systéme Nerveux de L'Homme et des Vertebrés. Maloine, Paris

Golgi C (1873) Sulla sostenza grigia del cervello. Gaz Med Ital Lombardia 6:244–246

Colby DE (1970) Induced estrus and timed pregnancy in cats. Lab Animal Care 20:1075–1080

Marín-Padilla M (1967) Number and distribution of the apical dendritic spines of the layer V pyramidal neurons in man. J Comp Neurol 131:475–490

Marín-Padilla M (1968) Cortical axo-spinodendritic synapses in man. A Golgi study. Brain Res 8:196–200

Marín-Padilla M (1969) Origin of pericellular baskets of the pyramidal cells of the human motor cortex. Brain Res 14:633–646

Marín-Padilla M (1970) Prenatal and early postnatal ontogenesis of the human motor cortex. A Golgi study. Brain Res 23:167–183

Marín-Padilla M (1971) Early prenatal ontogenesis of the cerebral cortex (neocortex) of the cat (Felix domestica). A Golgi study. Part I. The primordial neocortical organization. Z Anat Entzwickl-Gesch 134:117–145

Marín-Padilla M (1972a) Prenatal ontogenetic history of the principal neurons of the neocortex of the cat. A Golgi study. II. Developmental differences and their significance. Z Anat Entzwickl-Gesch 136:125–142

Marín-Padilla M (1972b) Structural abnormalities of the cerebral cortex in human chromosomal aberrations. A Golgi study. Brain Res 44:625–629

Marín-Padilla M (1974) Three-dimensional reconstruction of the pericellular baskets of the motor (area 4) and visual (area 17) areas of the human cerebral cortex. A Golgi study. Z Anat Entzwickl-Gesch 144:123–135

Marín-Padilla M (1978a) The Golgi method. In: Adelman G (ed) Encyclopedia of neurosciences, vol. 1. Birkhäuser, Boston

Marín-Padilla M (1978b) Dual origin of the mammalian neocortex and evolution of the cortical plate. Anat Embryol 152:109–126

Marín-Padilla M (1985a) Neurogenesis of the climbing fibers in the human cerebellum. A Golgi study. J Comp Neurol 235:82–96

Marín-Padilla M (1985b) Early vascularization of the embryonic cerebral cortex. Golgi and electron microscopic studies. J Comp Neurol 241:237–249

Marín-Padilla M (1990) Three-dimensional structural organization of layer I of the human cerebral cortex. A Golgi study. J Comp Neurol 229:89–105

Marín-Padilla M (1992) Ontogenesis of the pyramidal cell of the mammalian neocortex and developmental cytoarchitectonics: A unifying theory. J Comp Neurol 321:223–240

Marín-Padilla M (1995) Prenatal development of fibrous (white matter), protoplasmic (gray matter), and layer 1 astrocytes in the human cerebral cortex. A Golgi study. J Comp Neurol 358:1–19

Marín-Padilla M (1996) Developmental neuropathology and impact of perinatal brain damage. I. Hemorrhagic lesions of the neocortex. J Neuropath Exp Neurol 55:746–762

Marín-Padilla M (1997) Developmental neuropathology and impact of perinatal brain damage. II. White matter lesion of the neocortex. J Neuropath Exp Neurol 56:219–235

Marín-Padilla M (1998) Cajal Retzius cell and the development of the neocortex. Trends Neurosci 21:64–71

Marín-Padilla M (1999) Developmental neuropathology and impact of perinatal brain damage. III. Gray matter lesions of the neocortex. J Neuropath Exp Neurol 58:407–429

Marín-Padilla M (2010) Neuropatología evolutiva del cerebro dañado en las crisis epilépticas. In: Rufo M (ed) Las Crisis Epilépticas en los Desórdenes del Desarrollo Cortical. Madrid. Momento Medico

Marín-Padilla M, Marín-Padilla T (1982) Origin, prenatal development and structural organization of layer I of the human cerebral (motor) cortex. A Golgi study. Anat Embryol 164:161–206

Marín-Padilla M, Stibitz G (1974) Three-dimensional reconstruction of the baskets of the human motor cortex. Brain Res 70:511–514

Valverde F (1970) The Golgi method. A tool for comparative structural analysis. In: Nauta WJH, Ebbesson SOE (eds) Contemporary research methods in neuroscience. Springer, Berlin

Index

A
Applicability, 139, 140, 143
Ascending
    developmental incorporation, 76
    function maturation, 21, 23–28, 37, 38, 40, 41, 45–48, 130–135
    stratification, 11, 24, 26–28, 31–32, 134

B
Basket cell, 68–74, 77, 83
Brain
    development, 2–4
    vascular system, 87–90, 96–98, 100

C
Cajal, 139–143
Cajal Institute, 3
Cajal–Retzius (C–R) cell, 51–66
Cajal's old Golgi preparations, 3
Capillary growth, 99
Cat brain, 127, 128, 131, 136, 137
Chandelier cells, 68, 79–84
Cortical stratification (lamination), 1
C–R cell. *See* Cajal–Retzius cell

D
Descending function, 46–49
Development, 51–66, 87–100, 103–112, 127–137
Developmental cytoarchitectonics, 115–121
Double-bouquet, 68, 71, 77–79, 83–84

E
Expanding functional territory, 61, 65
External glia limiting membrane (EGLM), 103–106, 108–111
Extrinsic microvascular (Virchows–Robin) compartment, 87, 90, 92–97

F
First lamina, 7–10
First lamina astrocytes, 103–112
Functional transition (15th week of gestation), 22, 23, 24, 25

G
Golgi, 139–143
Gray matter protoplasmic astrocytes, 103–112

H
Human brain, 11–16, 18, 32
Human brain capabilities, 121

I
Inhibitory neurons, 67, 68, 72, 77, 83
Intrinsic microvascular (blood brain barrier) compartment, 94, 95, 98

M
Mammalian
    cerebral cortex evolution, 115, 116, 118
    embryos, 5–7, 9
    innovation, 37, 47
Martinotti cells, 68, 75–77, 84
Meningeal compartment, 88, 89, 93, 96, 97
Misconceptions, 139, 140
Morphology, 37–43, 45, 103, 106, 110
Multipolar morphology, 60–61, 63, 64

N
New nomenclature, 115–121
Number pyramidal cell strata, 115–120, 135–137

P
PCP. *see* Pyramidal cell plate
Persistence, 58, 59
Pial capillary plexus compartment, 89–93, 96, 97
Prenatal development, 11, 12, 26, 31–32
Primordial cortical organization, 5, 9–10
Protoplasmic astrocytes ascending cortical incorporation, 103
Pyramidal cell plate (PCP), 11, 17–28, 30, 31
Pyramidal neuron, 37–49
Pyramidal strata P6 and P7, 123, 124

R
Rapid Golgi procedure, 139, 140

S
Spatial orientation, 64, 65, 68–72, 84
Subplate (SP) zone, 7–9
Synaptic contacts, 39, 41–45, 47, 48

T
Tangential views, 59, 62–64

V
Vascular cortical perforation, 89, 90, 92

About The Author

Dr. Miguel Marín-Padilla, born in Spain, is Professor Emeritus of Pathology and of Pediatrics at Dartmouth Medical School (Hanover, New Hampshire, USA). He obtained his Medical Degree from Granada University in Spain in 1955 but subsequently emigrated to the USA, where he completed his clinical internship and several years of pathology residencies (Mallory Institute of Pathology, Boston), with emphasis on developmental and pediatric pathology. Between 1960 and 1962 he was a Teaching Fellow in Pathology at both Boston and Harvard Medical Schools. Dr. Marín-Padilla then pursued an academic career, ultimately becoming Full Professor in both Pathology and Pediatrics at Dartmouth Medical School. His research efforts have resulted in 180 publications, including abstracts, peer review papers, collaborative studies, book chapters, and this book. His developmental studies are recognized the world over and he has received various honors for his teaching and his research efforts. The most notable among these include: Alpha Omega Alpha (USA Honor Medical Society, 1981), Best Teacher Award (1988, 1996) and Faculty Speaker at Commencements (1985, 1991) at Dartmouth Medical School, Cajal Medal (USA Cajal Club, 1990), Gold Medal 'Aureliano Maestre de San Juan' University of Granada, Spain (1997), Honorary Member of the Spanish Neurology (1987), Pediatric Neurology (2000) and Neurosciences (2009) Societies, The Jacod Javist Neuroscience Investigator Award (USA, 1989), Honorary Member of the Royal Academy of Medicine (Murcia, 2001), the Gold Medal of the Community of Murcia (2001) and The Miguel Marín-Padilla Award for Excellence in Pathology (2006), an Annual lectureship at Dartmouth Medical School.

Printing and Binding: Stürtz GmbH, Würzburg